All about Locks and Locksmithing

All about Locks and Locksmithing

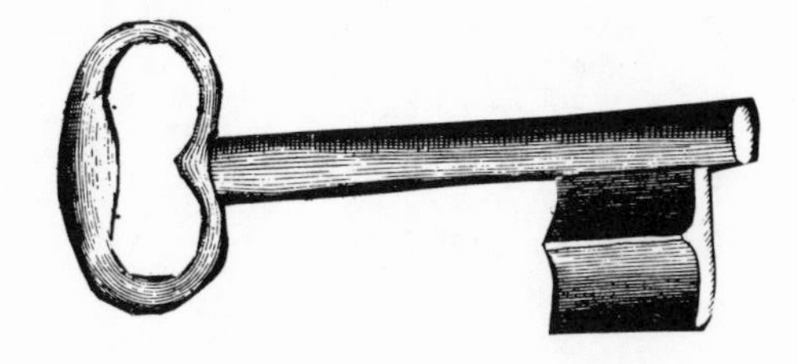

BY MAX ALTH

Hawthorn Books, Inc.
Publishers
NEW YORK

ALL ABOUT LOCKS AND LOCKSMITHING

 This book was manufactured in the United States of America and published simultaneously in Canada by Prentice-Hall of Canada, Limited, 1870 Birchmount Road, Scarborough, Ontario.

Library of Congress Catalog Card Number: 70-179105
ISBN: 0-8015-0150-4

DESIGNED BY ELLEN E. GAL

4 5 6 7 8 9 10

Foreword

Locks are puzzles, little machines, footnotes to history, and weapons with which we fight crime.

Some persons see the history of mankind when they examine a lock, its workmanship and sophistication bearing witness to our growing skill and cleverness. Others find pleasure and satisfaction studying, collecting, and displaying both locks and keys. Still others make locks their profession. Throughout history men have attempted to construct more beautiful and intricate locks, and some have founded fortunes inventing and manufacturing more secure locks.

Today security is of great concern to many people, and locks are an important means of obtaining security. In this book I discuss the relative strengths and weaknesses of the most commonly used locks —from the simple warded lock to complex electronic systems. To give the reader perspective, I have also provided a history of locks.

At the end of the book is a glossary of terms. This glossary should be consulted when you encounter terms with which you are unfamiliar. Some of the definitions may differ from what you expect or from what you have heard or read elsewhere. Although the locksmithing profession is now in its fifth millennium, practitioners of the art have yet to agree on a common terminology.

For those who collect and those who think they might want to collect locks and keys, I have devoted the opening chapter to some tips on collecting these puzzling little mementos of history.

M. A.

Contents

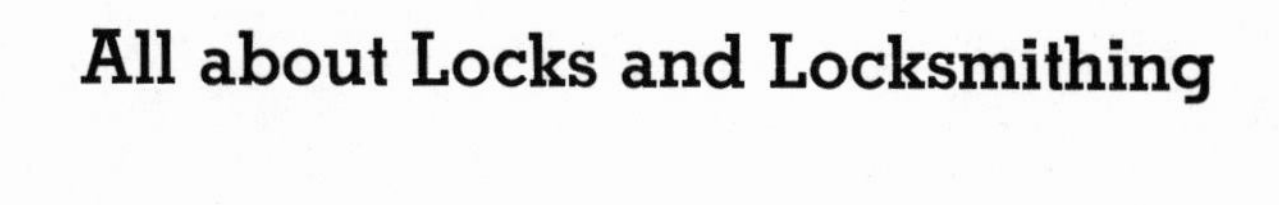

All about Locks and Locksmithing

Chapter 1

Why Collect?

Everyone who has a collection has been asked more than once: "Why collect?" Presumably each individual answers differently, yet there must be one or more reasons that are common to all collectors. Otherwise, they would not form societies, publish books and periodicals, or take delight in the collections of others.

The author of one book on collecting finds that as a group, collectors are more patient, more interesting, and more interested than noncollectors. He has found, for example, that dealers in antiques enjoy meeting with prospective buyers and take pleasure in discussing their wares. In a world crammed full of hucksters, it is a joy to meet with and talk to someone not aggressively selling himself or his product. In this respect collectors are not too different from tennis players and fishermen and electronics enthusiasts. Meet anyone who is interested in something other than himself, and you will find a person a bit out of the ordinary, someone who contributes to the life of others, if only in a small measure.

Collecting is a hobby—and sometimes a profession—that brings individuals into contact with other individuals with similar interests and, to some degree, similar depths of perception and awareness. Many lifelong friendships and even marriages began with little more than a common interest shared.

Like all other hobbies, collecting provides modern man with the necessary, even vital change of pace that enables him to survive willy-nilly in this mad, frenetic, fear-filled world.

THE ECONOMICS OF COLLECTING

A collection is something one displays with pride and discusses with authority. It is something tangible, something one can leave to the children or donate to a museum. When assembled with care and intelligence, a collection is usually an excellent investment. This holds true for seashells, coins, stamps, antiques, paintings, lepidoptera, and locks and keys.

Collectibles grow scarcer by the hour while the number of active collectors continues to grow. There is nary a cupboard or attic that hasn't been searched clean, and collectibles continue to be lost to fire and accidental damage. At the same time, however, population steadily increases. Demographic experts maintain that by the year 2000, the U.S. population will reach three hundred million. In the world at large another billion souls are expected to join us in a like period, bringing the total head count to four billion. Assuming the growth curve of collecting interest flattens, there will still be three collectors on the beach looking for exotic shells where today there are but two; three bidders at art auctions in place of two; three customers in the antique shop in place of two; and so on. The sheer weight of increasing numbers will hold the real value of collectibles reasonably steady provided there is no percentage increase in the number of individuals collecting.

While the real dollar value of collectibles can be expected to remain steady despite any presently unforeseen drop in interest, the numerical value of collectibles in dollars must increase. For a long time we have been subjected to a steady 6 per cent per annum rate of inflation. If this rate continues, and most economists believe it will, all material goods—antiques and collectibles included—can be expected to increase in numerical dollar value by at least 180 per cent between now and the end of the century.

On this basis alone, collectibles are on a par with the average stock or bond as a hedge against inflation. If, however, history repeats itself just a little, collectibles can be a more desirable investment than the average stock or bond.

The upward surge in prices paid for old paintings has been recounted too many times to be repeated again here. Let us instead examine some figures compiled by Dr. Richard H. Rush on the prices of antique furniture through the years. Dr. Rush, a prominent economist, financier, and former corporation president, is considered by many to be a leading authority on the price of art and antique objects. This writer strongly recommends his book, *Art as an Investment,* to all those seriously considering collecting.

For ease of computation, Dr. Rush used only Louis XV, Louis XVI, and old English furniture prices to produce the following chart, which reflects simple averaged prices for the years listed. 1925 is used as the base year.

1925	$100
1926	146
1929	193
1932	119
1939	47
1942	69
1945	172
1950	126
1955	167
1960	356
1965	570

Note that antique furniture prices were not immediately affected by the Great Depression that began in 1929, and that it was not until 1939 that prices fell a little more than half. Most likely this fall was due to the influx into America of important holdings by Europeans fleeing the Nazi holocaust. Note that when the war ended in 1945, prices were back up, whereas many of the stocks that fell on that Black Friday in 1929 dropped completely out of sight and were never seen again.

COLLECTIBLES AS TIES TO HISTORY

When discussing the value of a collection, it is not sufficient to consider its dollar equivalent alone; one must also consider the intrinsic value of the collection. In this writer's opinion the true collector selects and annotates his materials not just for their value as weapons against inflation or for their value as entree to special groups or as conversation pieces and ornaments. For this collector, the soul of collecting is contact with the past.

Today, more than yesterday and much more than the day before, we live in a world of change. Everywhere old buildings are being torn down. There is hardly a hamlet that hasn't been ripped asunder by urban renewal. The old, the familiar are rapidly disappearing to be replaced by possibly more efficient but definitely less human structures—very often of gigantic size. In New York City—southwest Manhattan—for example, the World Trade Center is under construction. There will be two towers. Each will cover about one acre and each will be 110 floors tall. The Trade Center has destroyed old Cortlandt Street and its environs. This was the mecca for radio buffs, where one could purchase Edison cylinder records in an unused state and 1924 magnetic speakers, where interesting restaurants still occupied Civil War and pre–Revolutionary War structures. Improvement? In a sense yes, but nothing has been left for the individuals who were born there, lived there, or merely knew the locale intimately.

New homes, new apartments, new cities and towns, are "new." They may be good or bad, tasteful structures or horrible "strip cities"—endless rows of garish facades along one boulevard. They may be lovely, with gardens and pools, or, more often, giant, cold warrens housing more people than London in 1800. No matter—there is nothing authentically "old" in their construction.

It is because of this constant bombardment of what is "new" that so many young and old are turning to the past, seeking some physical evidence that they are not the "perfect" generation. They wish corporeal reassurance they did not just happen. They wish to uncover ties with the past; they seek a sense of continuity with yesterday.

All collecting, but especially the collecting of artifacts, brings with it a sense of oneness with times past. And it is this altogether human desire to hold our shifting world still and in clear focus for a few moments that is at once the driving force behind collecting and its major satisfaction.

True collecting broadens one's view of the world. A person can acquire on the strength of dollars alone, and one can acquire lovely and valuable objects on the advice of others; but if the collection was not born of one's own love and sweat, it is not a true collection but merely a gathering of material objects. If the owner of these objects wishes to take true pride in his handiwork, he should point to the skill, acumen, and effort he invested in earning the money that purchased the collection. He may be a wealthy man but not a collector.

Learning the nature and history of the substance of one's collection; learning to properly evaluate, to select and not merely amass; meeting with others interested in the same discipline; discussing; reading; above all, learning—true collecting means all of these. Learning, as we know, is a growth process, and if old age cannot be avoided, growth by learning can at least deter it for a little while.

COLLECTING LOCKS AND KEYS

The collecting of locks and keys is essentially a journey into the past. Man has constructed and used locks and keys for more than four millennia, and although few of us can hope to own a lock or key from earliest times, there are sufficient examples of the locksmith's art from times between to provide all prospective collectors in this field with the chance of securing at least one piece of historical interest and excitement.

Locks and keys represent the efforts of both artists and craftsmen, often combined in one individual. Unlike early mason jars, for example, locks and keys were often designed for display as well as utilitarian purposes. Many locks and keys are ornamental. Frequently the collector owning a working piece will install it in his home.

From a financial point of view, locks and keys are also an excellent investment. Along with the tremendous surging interest in all things connected with the immediate and distant past has come an increased interest in things mechanical, and locks and keys are in actuality little machines. The immediate result is that the general price for locks and keys is well below the prices of items of similar age but which caught the antique buyer's eye earlier. To mention but a few of the items in this category, we have clocks, phonographs, spinning wheels, old electric motors, old toasters, and old toy locomotives.

This writer recently purchased a brass front-door lockset comprising handle, thumb lever, lock, plates, knobs, and interior trim for less than five dollars. When the patina of years has been removed and new keys have been fashioned, the lock will be installed in the front door. The age of the set is well over fifty years. An equivalent, brand-new front-door lockset of brass costs ten times more.

Cast-iron door locks, albeit somewhat begrimed with paint, of the simple warded type that mount on the inside of the door can be purchased for less than two dollars. Locks of this type may be up to seventy-five years in age. Frequently they are sold with the original iron doorknobs still in place. The same shops sell porcelain doorknobs of the same era for use with the same lock for about ten dollars. Why the vast difference in price? Collectors have taken fancy old doorknobs to their hearts; there still isn't a great deal of interest in the locks that accompany the knobs.

Further evidence of the public's comparatively minor interest in locks and keys—a situation that will not long remain—may be found in the scarcity of locks and keys carried by antiques dealers in general and the absence of dealers specializing in these items. New York City probably has three hundred antiques dealers on Third Avenue alone. Not one is primarily concerned with locks and keys, though most of them feature one item or another: clocks, furniture, ceramics, and even wooden Indians.

One fair source of locks and some keys is the urban renewal program. Some of the house-wrecking companies collect memorabilia, and these concerns may be contacted for locks and other items found in the demolished buildings. Most house wreckers are too interested in the rapid completion of the work at hand to save anything, but on occasion they sell old lumber and building parts at the site of their operations. These are not regular sales, as might be imagined, but can sometimes be located by following newspaper reports of urban renewal progress and new highway construction.

Locks and keys can sometimes be removed from condemned buildings before they are torn down. To do so one should obtain permission. Illegal entry, even with the honorable intention of preserving the past, is still a criminal act. Sometimes the watchman is a student of history too and a search of the premises can be made upon securing his goodwill.

For really old pieces at bargain prices one must search: There is no known alternative. Antique shops, fairs, flea markets, thrift shops, Salvation Army shops, rummage sales—all these are potential sources of interesting and even valuable examples of the locksmith's art.

How does one identify types of locks and keys? Some general information may be gleaned from the short history of locks in the following chapter. Each civilization produced its own distinctive type of locks and keys. Relatively recent locks and keys can sometimes be placed chronologically by examining their workmanship. Mass production of locks did not begin until well past the 1850s. Locks prior to this time bear evidence of handwork. Handwork can be recognized in several ways. For one thing, the nuts and bolts of one handmade lock usually are not interchangeable with nuts and bolts taken from another lock. Early craftsmen cut their own threads. Parts too are usually not interchangeable, and the farther one goes back in time, the greater the difference in part size and overall lock design, although locks from different shops may be alike in general appearance and design.

Handmade parts usually do not have the square edges of machine-made parts. Also, handmade locks will usually have more pieces. Without modern machine tools it is difficult to fabricate a complex shape from one piece of metal. It is easier for the craftsman to assemble a shape from a number of pieces. And as we go farther and farther back in time, handwork becomes cruder, rougher. In the early 1900s the statement "He makes it just as good as a machine" was high praise. For this reason and for personal satisfaction, many locks produced during the interim period when hand manufacture ceased and machine production began were carefully finished with machine-like precision. To the novice these locks may appear as though they were mass-produced, but in time one learns to recognize the touch of the artist.

Chapter 2

The Lock and Key in History

Time has long dismissed the memory of the man who invented the lock, but with some degree of accuracy we can reconstruct the first lock and its time in history. Without doubt it was based on a key, which in our language still means "something that gives an explanation or provides a solution."

The time is the beginning of the Neolithic age. Man has as yet not discovered metal, but is busy domesticating his animals and learning how to farm the land. Some experts believe the Neolithic age began as early as eleven thousand years ago, when the glaciers began their last retreat and the earth was subjected to great climatic change.

The place of the first lock is probably the hills of Kurdistan, north of the Tigris-Euphrates river valley—the Fertile Crescent—where civilization appears to have begun in the Western world. Here, records of man's first definite movement out of the caves appear. Here the oldest sickle blades of flint for harvesting and pounding-stones for milling and grinding grain have been found.

Until this time in his history man had been largely a hunter. He possessed nothing larger than he could carry and nothing so important he could not bear to leave it behind. Now he was a farmer. He had hard grain that could sustain him and his tribe from one crop to the next. Now he had riches and he was troubled. The grain was too bulky to be carried about as he tended his fields and too valuable to be left unguarded. There were wild beasts and an occasional marauder from the harsh and barren Anatolian plateau north and west of him.

We can see him in our mind's eye, straining his faculties for a solution to his dilemma. We cannot know how or when the answer came, but it came in the form of a key. The key was first, and it was a thing of magic and mystery, as were to be all keys well into the eighteenth century.

Having invented the key and having carefully tried it in secret several times, the world's first locksmith gathers his little clan around himself. He swells up with importance and declaims, "See the entrance to the cave. Now it is open and free to the wild creatures and other enemies. Soon I shall perform the magic that shall seal it tightly and safely against all evil. But first all must leave. This magic cannot be accomplished in the presence of others. I must be alone."

Muttering to themselves, shaking their long-haired heads, the little group departs. Quickly the inventor-magician rolls a large, smooth stone tightly against the cave opening. Looking over his shoulder to make certain no one is watching, he scoops a small hole in the dirt beneath the front of the rock. Then he wedges his "key," a shard of stone, beneath the boulder and covers it with soil and leaves. When he calls his people back, the earth bears its original appearance.

The strongest in his tribe cannot gain entrance to the cave. They do not have the "key." They do not know the secret. As we shall see, the secret of a lock's operation is its source of strength. This has been so ever since the lock was invented and remains even today the single greatest barrier to unlawful entry.

THE OLDEST LOCK EXTANT

Paul Emile Botta, credited as the father of modern scientific archaeological investigation, was the French consul at Mosul, in Persia, in 1842, when he discovered what is believed to be the oldest lock extant. Digging at Khorsabad, the ancient capital of Persia, Botta uncovered the remains of the palace of Emperor Sargon II, who ruled the region from 722 to 705 B.C. Sargon modestly called his palace Sargonsburg. It covered twenty-five acres of the one-mile-square Khorsabad, which, it was estimated, could comfortably house eighty thousand persons in its prime.

Working slowly, Botta's men uncovered a large, ornate hall of obvious importance. At one end stood two tall Assyrian statues. Typically each bore the head of a man on the body of a bull equipped with the wings of an eagle. Behind the pair of stone bulls was a single heavy gate of wood locked in place by an equally massive lock of wood.

The lock was on the inside of the gate. There was no keyhole in the sense we know it today. Instead there was an arm-sized hole cut through the gate at one side of the lock. The heavy wooden lock had a heavy wooden key, so large that a child could not carry it. To operate the lock one had to pass the entire key through the hole and match the pegs mounted on the key to smaller holes cut into the square wooden bolt. If the correct key was used, the pegs fitted the holes and the key became a handle with which one could slide the bolt free and open the door.

Large "city keys" of this size were symbols of authority and were carried importantly about on the shoulder. Even today the "key to the city" is often given to an important guest as a sign of public welcome.

There is no record of Sargon's locksmith, but we know that the Sargon lock was not invented during the emperor's lifetime. Columns of the great Egyptian temple at Karnak, constructed about 2000 B.C., picture this type of lock in bas-relief. Thus the Sargon lock was at least 1,300 years old in its design. Elsewhere in Egypt somewhat similar locks appear in wall paintings. Similar keys, some thirteen to fourteen inches long, for use with doors to private homes have been found buried with the dead at Luxor. Most of the keys are of teak, a hardwood, with pins of iron or wood. Some have ivory handles and are inlaid with gold and silver.

Why only keys were buried with their owners and not the locks is a matter of speculation. Possibly the keys were believed necessary to unlock doors in the future world. Perhaps the practice derives from the same superstition that led men of much later eras to hold keys sacred and magical, but rarely locks. Perhaps the practice is due to the archaic tendency of man to denigrate women: The key, symbolizing the male, is worthy of honor, whereas the lock, symbolizing the female, is not. We do not know for certain.

Although the Sargon lock is the oldest known lock in existence, it is far from a unique mechanism. Locks of this same general type and based on the same principle have been found over most of the Western world—in England, Scotland, Germany, France—and as far east as Japan and the East Indies. Wooden locks were in common use in Egypt and Turkey as late as the beginning of the twentieth century and may still be found in daily use in some remote parts of Turkey and the Near East.

Egyptian-type wooden locks have been used in Cornwall, England, from time immemorial. They were probably introduced by the Phoenicians who came to Cornwall in search of tin, the element that converts copper to bronze. All the Celtic nations used locks of wood. An example may be found in the Museum of Antiquaries in Edinburgh. One writer reports wooden locks common in the highlands of Scotland well into the 1800s. Carved from native oak and yew, the wooden lock was far more resistant to British weather than the lock of iron, and far less costly too. Locks of wood are still used on many shelter huts in the Alps.

CONTRIBUTIONS OF THE GREEKS AND ROMANS

Greek locks followed Egyptian locks chronologically, but not scientifically, for Greek locks were far simpler than their Egyptian forebears. However, the Greeks are credited with inventing the keyhole, eliminating the need to poke one's entire arm through a hole in the door to reach the lock inside.

Pliny the Elder is famous for his *Natural History,* an encyclopedia containing twenty thousand facts, which he proudly presented to Emperor Titus. In this great work Pliny tells us that the key was invented by Theodorus of Samos in the sixth century B.C., over seven hundred years before Pliny compiled his encyclopedia. Homer, however, writing or singing some 250 years prior to Theodorus, established the presence of Greek keys. In Book XXII of *The Odyssey* he writes: "Penelope took a crooked key in her firm hand, a goodly key of bronze, having an ivory handle. She loosed the strap, thrust in the key, and with a careful aim shot back the door bolts. As a bull roars when feeding in the field, so roared the goodly door [to Ulysses' treasure chamber] and flew open before her."

We cannot dismiss Pliny the Elder's statement as pure conjec-

1. Model of an Egyptian lock of wood. One tumbler has been removed from its well to show the operation of the key. The horizontal bar is moved to one side by hand after all three pins are lifted. *(Courtesy Eaton, Yale & Towne antique-lock collection)*

ture, however. Samos at the time of Theodorus, Herodotus tells us, was one of the finest cities in the world, the very center of flowering Ionian civilization. Theodorus was an architect and sculptor. His father, Rhoecus, invented casting in bronze. Certainly there was wealth aplenty in need of protection in that fair city, and Greek artist-craftsmen of that time and later are famous for their creations in metal. Among their remaining artifacts is an early computer used as an aid to navigation. It is possible that Pliny the Elder did not err but that our interpretation of his statement is wrong. Perhaps he meant that Theodorus invented the first *cast bronze* key, and nothing more.

Greek locks were simple by modern and Egyptian standards. Their keys were merely large hooks shaped like a sickle. They were so large they had to be carried over the shoulder, and perhaps the security of the Greek locks lay in their very size. A long, strong curved piece of metal was necessary to reach through the small keyhole and slide the dead bolt back from the jamb. Metal was scarce and expensive. Individuals poor enough to want to steal did not have the price of the metal necessary for the key.

The Greeks also used lengths of rope tied into complicated knots to hold their doors fast against intrusion. The famous Gordian knot of Phrygia, which none could untie, is the best-known example of this type of Grecian lock. Alexander the Great didn't have time to fuss with the problem; he simply cut the knot with his sword. Had he been able to untie it, he might have won even greater fame.

The knot couldn't be untied because it was formed of a continuous length of rope: First the knot was made, then the ends of the rope were woven together so that the point of juncture could not be easily seen. Pushing and pulling on the rope would just make it move around in a continuous, complicated circle.

Contrary to popular opinion, the Romans of antiquity were not merely copiers. They took the simple Greek lock and its three-foot-long key and replaced them with a more complicated lock and a key not too much larger or different from what we know today.

The Romans were the first to use springs in their locks; the first to introduce wards or obstructions in the path of the key. The Roman locks of old were much the same in principle as our warded locks. Most of the Roman keys had to be turned to operate the lock mechanism, but some had merely to be inserted to open the lock. There are fairly modern Chinese locks that work this way.

The outer doors to Roman homes were made of wood. They turned on pivots set in the lintel and threshold (top and bottom of the doorway). Door locks, other than padlocks, were integral with the door itself. Usually there was a metal plate with a keyhole fastened by rivets to the front of the door. The lock mechanism was held in channels cut into the door frame behind the plate.

Rome was mistress of the Western world for more than a thousand years. During this period her artificers improved on the Greek lock while her statesmen, generals, and merchants popularized it. The Romans of antiquity had an unbridled passion for power as well as the means to power—hard cash. If we can assign to any particular people of the past the need for locks and keys, we can certainly assign that need to the Romans. At the zenith of Rome's glory, she must have had more locks than slaves, and slaves she had aplenty. It is therefore surprising that very few Roman locks remain today.

Two reasons for this scarcity are plausible. The Romans were an eminently practical people. They preferred iron over bronze as a lock material because iron was stronger and cheaper. That iron had a short life in the moist air of Rome would be of small moment to people who lived in an age that saw twenty-five years as one's expected life span.

The second reason for the scarcity of Roman locks may be the inconsiderate length of time it took Rome to decline and finally fall. Rome wasn't sacked in a day and forgotten but was systematically looted by a succession of invaders over many, many years. Those locks that weren't smashed when the barbarians forced their entry, and those padlocks that were not carried away still attached to the coffers they guarded, were stripped from doors and chests and carried home as splendid mechanical marvels of Roman civilization, then forgotten and permitted to rust into oblivion.

Roman keys, however, are an entirely different matter. There are antique Roman keys by the thousands in important collections all over the world. Some were collected by Renaissance princes, others were kept by simpler folk—the mystique of the key again.

Roman keys were made of bronze or iron and bronze. Since

Roman togas came without pockets, the smaller keys have rings which were placed on the fingers. Others have volutes for accepting a ribbon or chain. Some Roman keys are flat, somewhat like modern keys. Most have turned ends where they are to enter the lock. Some keys were made very small to fit the fingers of women. Others are large. They were carried loosely or on metal rings by household stewards.

LOCKS AND KEYS OF THE MIDDLE AGES AND THE RENAISSANCE

In the thousand-year span that followed the fall of Rome few technical advances were made in lock design. The simple warded lock remained the basic mechanism. For greater security locksmiths turned to complexity and deception.

Little is known of the locks and keys produced in Western Europe immediately following the fall of Rome, but after the sixth century two groups of locks and keys of a distinctive appearance were manufactured. A number of them remain in museums and private collections.

One group is called Merovingian after Merowig, who founded the Frankish dynasty that ruled in Europe from about 500 to 751. The second group of locks and keys is named after Carolus Magnus, better known as Charles the Great or Charlemagne, king of France (768 to 814) and emperor of the West (800 to 814). This group is called Carolingian.

The two groups of locks and keys are somewhat similar in appearance, and it takes familiarity before one can identify them with a degree of certainty. Generally the Carolingian-period keys have bows shaped somewhat like a bishop's miter. Usually they are flat and made of bronze. They look something like a chair in profile.

Merovingian keys are generally lighter in form and simpler in structure. Their shanks sometimes terminate in pins and their bits are usually less complicated than those of the keys manufactured by the Carolingian artisans who followed.

The first all-metal locks are attributed to English craftsmen working during the reign of Alfred the Great, who interested himself in fine locks. We do not know exactly when locksmiths on the continent began making locks entirely of metal, but it was probably not much after 900, when the reign of Alfred ended.

The start of the Gothic period (roughly from the middle of the twelfth century to the start of the fifteenth) finds the art of lock making considerably advanced from that of Roman times. Keys are still made of iron or bronze, but now they are all flat, with far more elaborate bows and bits. The early horizontal keyholes have been replaced with present-day vertical keyholes. This called for a more complicated internal lock mechanism, but Gothic craftsmen were up to it.

As the value of goods to be safeguarded rose, the respect and honor paid men capable of making locks and keys also increased. Charles IV of Germany created the title of master locksmith in 1411. Master clockmakers like Jorg Heusz and Hans Ehemann (reported to have invented the letter-ring combination lock), both of Nuremberg, and others of their skill were invited to make locks and keys for the royalty of France and Italy.

We find little dramatic change in the art of locksmithing with the advent of the Renaissance. With the exception of the self-locking bolt, no major mechanical improvements appeared. Previous locks required that the key be turned to lock the door as well as unlock it. The self-locking bolt has a beveled end and a spring. Close the door, and the bolt snaps into place.

Less and less work was done by the locksmith at his forge. More work was now done with saw, hammer, chisel, and file. Keys and locks lost their rough-hammered look and gained in sophistication and ornamentation. More and more effort was spent at concealment —hiding the keyhole behind an ornament or within a design. Keys and locks both became works of art, as desirable for their appearance as their utility. The most attractive are those made in France between 1550 and 1650. The keys are called *clefs de chef-d'oeuvre,* or masterpiece keys. They were too good for use but were put on display in the guildhall and in the atelier of the master locksmith.

It is generally contended that Italian keys did not equal their northern contemporaries. English, French, and German keys were

2. Early sixteenth-century lock *(Courtesy Eaton, Yale & Towne antique-lock collection)*

almost always cut from a single slab of metal. Italian keys were most often made by assembly of many carefully fitted pieces. It is this writer's contention that in the end, the Italian keys will prove the more interesting to collectors.

Spanish artisans of this era produced equally attractive locks and keys. In most points they are indistinguishable from those made in France and Germany. The Spanish artists paid considerably more attention to details, though. The exposed portions of the staples and bolts that held Spanish locks in place were often carved to represent figures and heads. Some therefore consider the Spanish locks to be the more complete works and more attractive.

Multiple keyholes and multiple locks were fairly common. One chest had a total of twelve bolts operated by several keys. Typically, when Princess Isabelle of Bavaria had an ultra-secure lock made to guard the apartments of her ladies-in-waiting, the lock required five keys to operate it.

When Henry II of France wished to keep his mistress to himself, he had three locks placed on the door leading to her rooms in the castle. Only his key could open all three locks. Three different keys, ostensibly used by three individuals, were necessary to open the locks. This is a very early example of master-keyed locks.

Some chests had hidden traps for the unwary. When opened, one chest presented a number of holes, purportedly for insertion of the fingers for lifting a tray. If the thief did so, a spring would catch his fingers and lock him to the chest.

THE ADVENT OF THE MODERN LOCK

Modern locks—the locks we know and use today—are the result of a gradual development that began with the Industrial Revolution. Until this general period of time, more effort was spent on decoration and concealment of the keyhole (there were even locks with smaller locks that kept their keyholes shut) than on improving the lock itself.

One of the first of the modern locks, patented in 1778, was invented by Robert Barron, of England. Like the Roman lock, it had wards around its keyhole, but in addition it had a pair of spring tumblers of different sizes which checked the bolt. To operate this lock one needed a key shaped to get past the wards and fitted with a bit that was cut so that it lifted the tumblers precisely out of the bolt. From a security point of view it was vastly superior to all other locks of its time.

In 1784 Joseph Bramah patented the first of his many lock inventions. He had already invented a machine for cutting quill pen points, a hydraulic press, and an improvement on the steam engine (which led him into a succession of lively lawsuits with the more famous steam-engine man, James Watt). Bramah was a brilliant production engineer. He reorganized lock-manufacturing, introduced the principle of mass production, and started the modern lock-making industry.

Perhaps it was coincidental, but Bramah's lock appeared at the crest of a sensational wave of burglaries. The Bramah lock was more complicated and "secure" than the Barron lock and was the first high-security lock that was operated by a key small enough to be carried conveniently.

Bramah was also a showman. He widely advertised his lock as being as "impregnable as the Rock of Gibraltar," and in 1811 he offered two hundred guineas (roughly two hundred thousand dollars today) "to the artist who can make an instrument that will pick or open this lock."

Shortly after Bramah made this daring offer, the British government offered one hundred pounds of their own to anyone who could devise a pick-proof lock. There is no record of whether or not the prize was ever paid.

In 1818, seven years after Bramah's prize offer, the three Chubb brothers, Charles, Jeremiah, and John, patented the first of their series of improved locks. Their lock was based on the lever tumbler principle.

Both the Bramah lock and the Chubb lock were on display at the 1851 International Industrial Exhibition in London. Using lock-picking tools of his own devising, an American, Alfred C. Hobbs, an expert locksmith and one of the best "legal" lock-pickers in the United States, opened both the Bramah and the Chubb locks within twenty-five minutes. Bramah, claiming he would stand by his now forty-five-year-old prize offer, requested a repeat performance. For this he brought forth a newer and larger lock. It took several days, but Hobbs succeeded again. Bramah protested because of the time length, but the two hundred guineas were paid.

Hobbs also succeeded in picking a Cotterill lock. Cotterill then added a "detector" to his lock and called it the Patent Climax-Detector lock. The detector lock had a weighted part that could easily be tipped into a "locked" position. Once the weight had been shifted by someone using the wrong key or attempting to pick the lock, only the correct key could open the lock. (Hobbs later improved on the Cotterill lock.)

Hobbs found England congenial. He founded his own lock and safe company there and stayed on for thirty years. A few years after he had started manufacturing his "improved" lock, one of his English workmen succeeded in picking a Hobbs lock. It took months of study and special tools, but he did it, vindicating once again the statement made by more than one lock expert, "If there is a keyhole, the lock can be picked."

In the Colonies, lock development trailed behind the technical level prevailing in Europe. The better early locks were imported. Indigenous locks were simple affairs, either wrought on the forge or fabricated by the cabinetmaker along with furniture and utility cabinets.

On the farm and in the simpler homes the latch string was used. A wooden bolt was installed horizontally on the door. One end fitted a hole cut in the jamb; the other was tied to a length of string. During the day the string ran through a hole in the door. A visitor could pull on the latch string and open the door (hence the saying, "Our latch string is always out for you"). At night the string was drawn back inside and there was no easy way of unlocking the door from the outside.

Few native craftsmen in these early days signed their locks. Those that did appear to have been geographically limited to Lebanon, in the Pennsylvania Dutch area of Pennsylvania, where two generations of Rohres made and signed their names to a series of German-type locks. In contrast, English locksmiths began signing their handiwork as early as 1700. The more famous include Joseph Key, who constructed a royal lock for St. James's Palace for eight

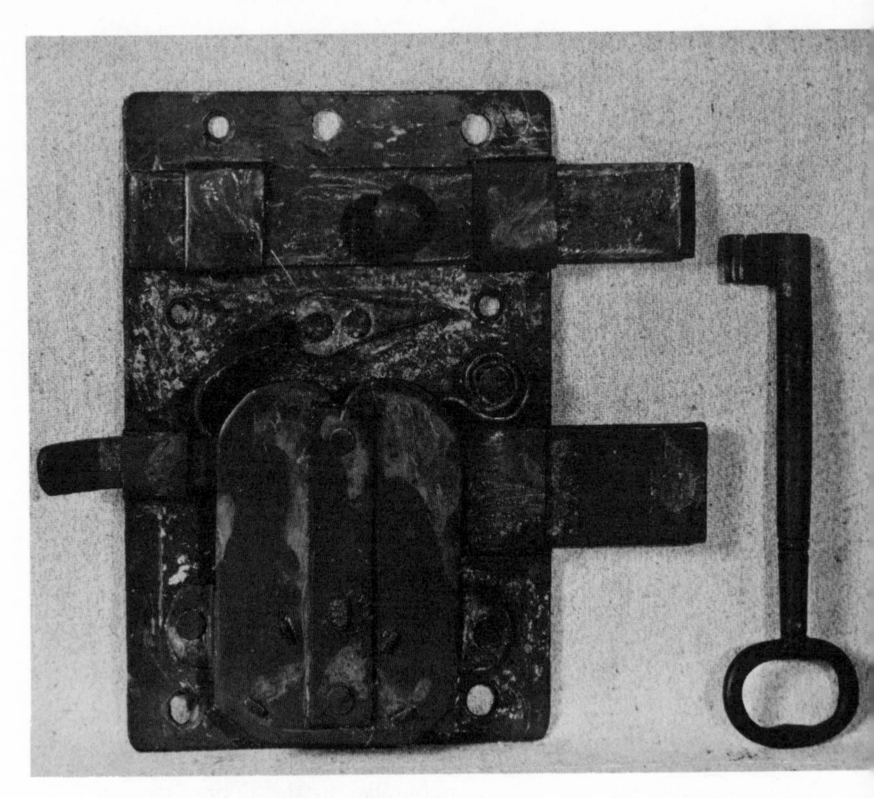

3. Early American hand-forged lock *(Courtesy Eaton, Yale & Towne antique-lock collection)*

hundred pounds; John Wilkes, who affected the signature *Johannes Wilkes de Birmingham fecit;* and another who signed himself *Philip Harris Londoni fecit.*

As industry in the United States progressed, American craftsmen copied the Barron, Bramah, Hobbs, and other British locks and to some degree improved upon them. At the start of 1836, Solomon Andrews, of Perth Amboy, New Jersey, made a lock that had adjustable tumblers and an adjustable key. If an unauthorized individual secured a duplicate key or if the owner of the lock wished to withdraw access from one to whom he had given a key, the lock owner could change the lock and change only his own key to match the new mechanism.

During this period J. Perkins, of Newbury, Massachusetts, patented a complicated bank lock with a key made of washers that could be rearranged when desired. This was another means of producing easily changed locks and keys.

In 1836 Day and Newell, of New York, improved on the Perkins lock by constructing a changeable lock with two complete sets of changeable tumblers. Like other proud lock inventors and manufacturers, they offered a five-hundred-dollar prize for picking it. The prize was promptly won by a local machinist, so Newell redesigned his lock and called it the Parautopic bank lock—from the Greek, meaning "hidden from sight." The Parautopic lock won a gold medal in 1847 from the National Mechanics' Institute of Lower Austria, no mean feat for an American product at any time.

In 1848 the Herring Company produced a lock that soon was called the grasshopper. The key to the lock was a semicircular, toothed piece of metal. To open the lock the key was inserted and the lock handle then turned. When the door opened, the key, jumping like a grasshopper, was vigorously ejected.

THE EGYPTIAN MECHANISM IS REINCARNATED

Linus Yale, Sr., appears to have made models of his now-famous pin tumbler lock well before he patented it in September, 1857. Hobbs described it in 1853 and admired it, though report has it that he was able to pick the Yale lock. "It is something like the Egyptian and something like the Bramah," he is supposed to have muttered.

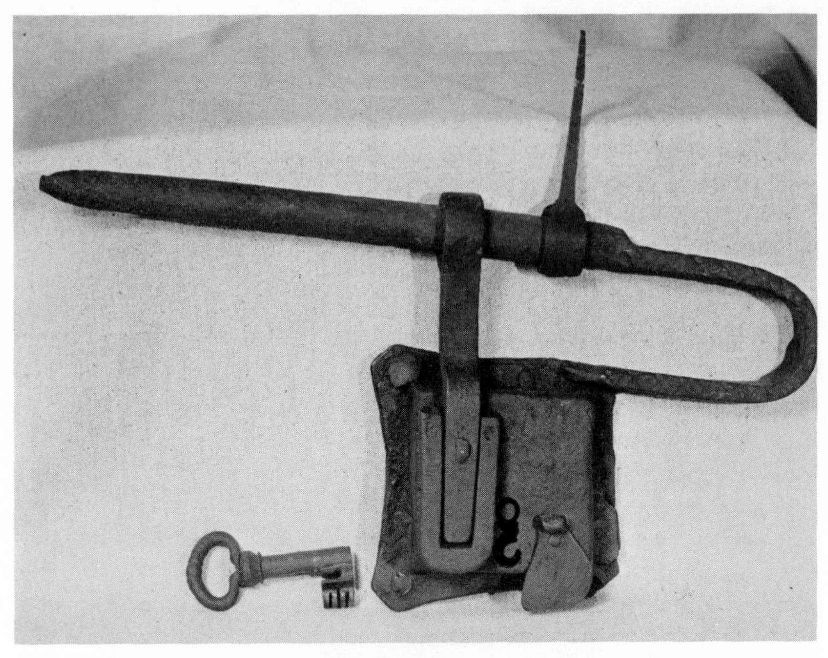

4. North European padlock from about the seventeenth century *(Courtesy Eaton, Yale & Towne antique-lock collection)*

The Yale lock had two cylinders, one within the other, held together by a series of pins.

Linus Yale, Jr., began as an artist but early turned to inventing and naming locks. His inventions include the Yale Infallible Bank Lock, the Yale Magic Bank Lock, the Yale Double Treasury Bank Lock, the Yale Monitor Bank Lock, the Yale Double Dial Bank Lock. Today his titles may seem grandiose, but the Yale lock with its improvements is the lock design most used in the world today, and Junior's Double Dial Bank Lock is the standard for all present-day combination locks.

The Yale pin tumbler mechanism has won worldwide approval because it is a very secure lock, it is fairly inexpensive to manufacture, it is easy to install, it is reasonably easy to change for a new set of keys, and its keys are small.

It is interesting to argue over the relative measures of security possible with the Bramah, lever tumbler, and pin tumbler locks, but experts maintain that the pin tumbler design is the best. Modern manufacturing skill permits the space beneath each pin to be cut to any of ten distinct heights. Commercial pin tumbler locks are made with anywhere from three to seven pins. In theory, that would produce 10^7—or as nonmathematicians say it, ten million—possible different keys for a seven-pin tumbler lock. In practice, however, only some thirty thousand to forty thousand different keys are

5. An original Yale lock, first of the commercial series, about 1865 *(Courtesy Eaton, Yale & Towne antique-lock collection)*

possible for any seven-pin tumbler lock. More pins are possible, but no locks having more than seven pins are commercially manufactured.

For this reason the pin tumbler lock is the one most used today where high-level security, but not the highest level possible, is desired. Bramah locks are still manufactured in England. Lever tumbler locks are still manufactured and used for bank safe-deposit boxes and in mental institutions. They are used in the latter because lever tumbler blanks are rare, while it is relatively easy to secure pin tumbler blanks and cut or have them cut to size.

For maximum security we depend on the work of James Sargent, who invented the magnetic lock in 1866 and perfected the time lock in 1873.

When combination locks first began to appear and bank vaults were themselves made stronger in the mid-1800s, crooks varied their tactics to suit. Instead of attempting to dynamite the vault open, they would kidnap the teller or bank president and force him to give them the combination.

When this was stopped by time locks, crooks adopted the practice of meeting the early-morning bank personnel and waiting with them for the time lock to open. In the present stage of the contest, the banks are countering with trained detectives disguised as early-morning bank customers, and they too now wait for the time mechanism to open.

Should you wonder what happens when the bank's clock fails, don't. In modern aerospace verbiage, they employ a redundancy of clocks. If one or two fail, there is always another present to carry on. So far, none have failed, but it will be interesting to those not directly concerned should the redundancy be insufficient.

Just a note on fakes. Locks and keys are only relatively less prone to "fakemanship" than art works. French Renaissance masterpiece keys were imitated in the nineteenth century to supply collectors. Because locks were more difficult to fabricate than keys, fewer of them were faked, but caution with an expensive item is advisable.

Chapter 3

Token Security: The Warded Lock

Warded locks are the simplest of all locks in use today. They provide the very minimum of burglar protection and are therefore only used for relatively minor token-security tasks such as locking luggage and shed doors. Warded locks are no longer installed on exterior doors, although many older homes and apartments still have these locks.

Generally, every paneled door in an old house, except the front and rear doors, carries a cast-iron warded lock. These locks are easily thirty-five or more years old, and may be as much as one hundred years old. Many small manufacturers have come and gone during their span of popularity. When greater security is wanted, common practice is to install a rim lock above the warded lock.

Parts for cast-iron warded locks can sometimes be found in large old hardware stores and at some locksmith shops. Warded locks and parts for warded locks can be found in junk shops, at the site of wrecked homes, and in cellars and attics, where entire doors are sometimes stored. Cast-iron warded locks can also be found in antique shops. No matter where found, do not expect any modern degree of part interchangeability. Carry your old lock with you and fit the parts right then and there. Do not use bolt hole placement and measurements alone to guide you in purchasing a new lock. Sometimes the screw holes match but the spindle holes do not, or everthing matches but the dead bolt is above the latch instead of below. Warded locks last a long time, and many changes have been made.

Warded padlocks, however, have very limited life spans. They should not be left to the ravages of weather because they will rust closed. And they should not be placed where they will be opened and closed every day; they wear out too rapidly. The so-called dollar lock (which like the five-cent candy bar is now probably several times its original price) will survive fewer than a thousand opening and closing cycles.

Warded locks derive their name from the word "ward," which means the action or process of guarding. "To ward off evil" is a common use of the word. The warded lock has wards, or guards, that protect against the entrance of an unauthorized key. The number of wards per lock varies, with two being the practical minimum; cast-iron door locks and low-price padlocks usually have two wards.

Warded locks are usually made of cast iron and in two parts. One part usually mounts and holds the mechanism, while the second part acts as a cover. In some of the lower-cost models the cover is a section of sheet metal. Warded locks are either mortised into the thickness of the door and held in place by screws through their front plate, or they are mounted on the surface of the door. In the latter arrangement the lock is fastened on the inside of the door, usually by long wood screws that pass through the case. The mortised lock is more desirable because it is neater and invisible. The surface lock takes up more space, which may preclude its use in a small closet, where one's hand can get caught between the knob and the lock. Generally, only the surface-mounted design lock is equipped with a safety latch (see Figure 6).

The safety latch generally locks the spring lever latch. When this is done, the door may be opened from the inside by turning the doorknob, but one must use a key to enter from the outside.

The weakness of the warded lock lies in the relative ease with which the obstructions can be circumvented and in the limited number of key changes (different key shapes) possible with the most complicated warded lock. The security of a lock lies in its strength and in the number of different keys possible with that type of lock. If, for example, a lock can be made to take no more than ten different keys—that is to say, ten different key changes—then one key out of every ten will fit that lock design. In an apartment house with one hundred tenants and one hundred locks, there will be ten peo-

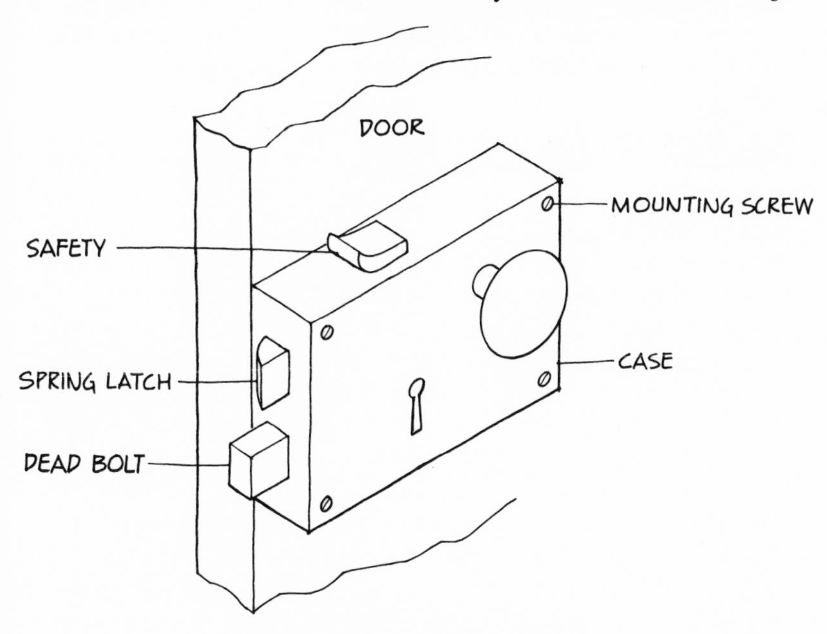

6. Typical surface-mounted warded lock

ple holding a key that will fit nine other locks than their own. Stated from another point of view, each tenant will have key access to nine apartments in addition to his own.

At best, the most secure and complex warded lock admits of no more than fifty different keys. This means that there is one chance in fifty that someone else's warded lock key may fit your lock. Actually, wear in both locks and keys reduces that number considerably. In practice, warded locks were and possibly still are manufactured with as few as four different keys to as many as fifty changes or different keys to a set. Usually, but not always, no key from one set fits a lock from another set.

And even though there may be fifty different keys to a set, any single key in that set can easily be altered to fit all the other locks in that set, and generally all the locks in all other sets.

Warded locks can be picked in a matter of seconds by an expert. The tool may be a bent nail, a slim allen wrench, or a strong safety pin. Even an unknowledgeable person can pick a lock of this type in an hour or so. How to pick these locks and make skeleton keys for them is discussed in Chapter 10.

PRINCIPLE OF OPERATION

The design principle behind the warded lock is straightforward and simple. Figures 7–9 illustrate the basic workings. In Figure 7 we see a typical keyhole and a key cut to fit this particular keyhole. Note the projection at the side of the keyhole and the corresponding slot on the side of the key's bit. As illustrated, this is not the only possible keyhole shape. In olden days when the warded lock was the best available, more complicated keyholes, a few of which are shown, were used.

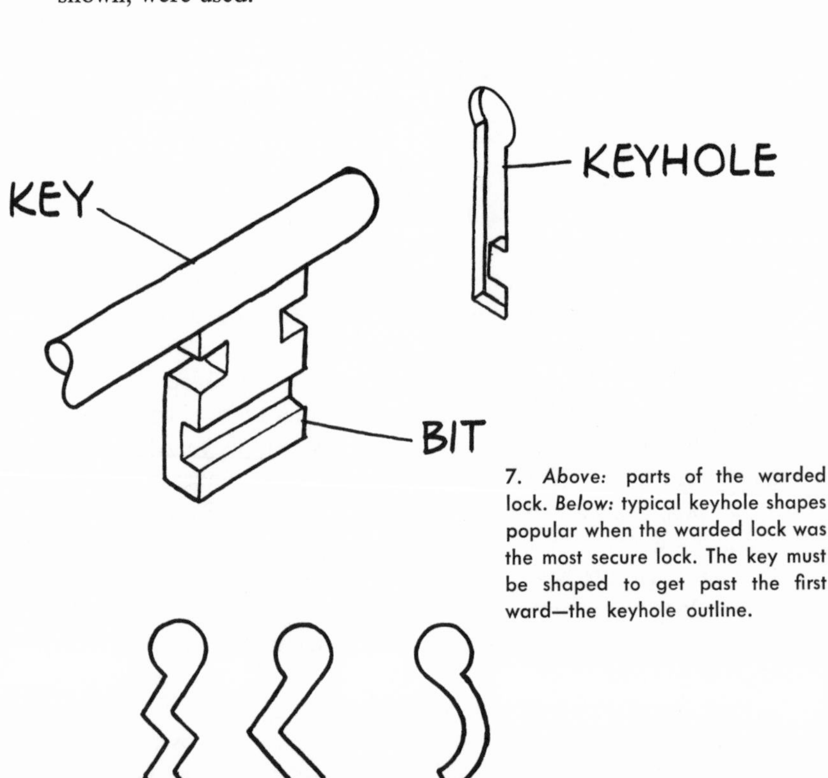

7. *Above:* parts of the warded lock. *Below:* typical keyhole shapes popular when the warded lock was the most secure lock. The key must be shaped to get past the first ward—the keyhole outline.

In Figure 8 the key has passed through the keyhole and is now in the process of being turned past one ward. Only one is shown. Note the bitting on the sides of the key necessary to accommodate the side wards. In Figure 9 the key has been turned past the side wards and now engages the dead bolt. In the design shown, turning the key moves the dead bolt directly.

The warded lock design illustrated is not the only possible design. There are other different, and possibly more complicated, arrangements of parts. In some locks the key does not activate the dead bolt directly but operates through a series of levers.

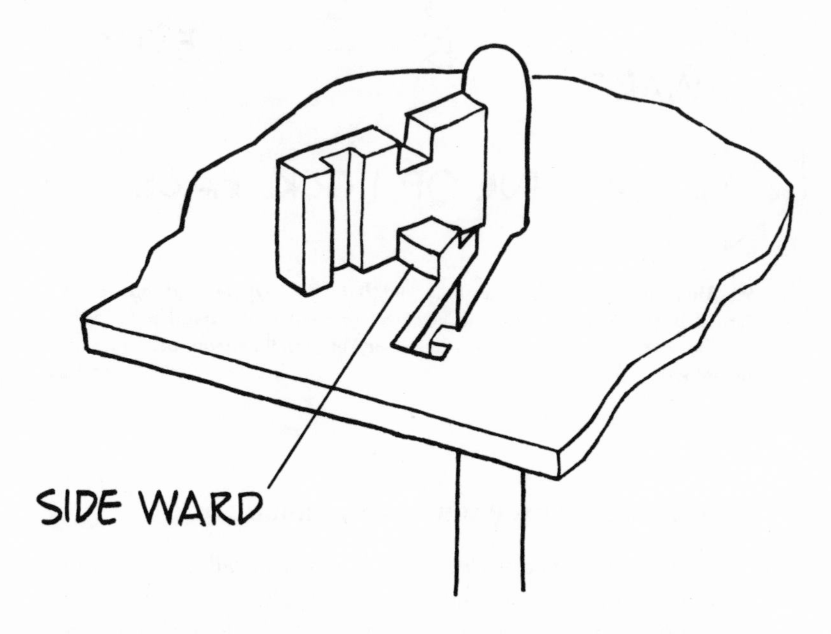

8. In order to be capable of turning, the side of the bit must be shaped (bitted) to match the wards on the side of the case. Only one is shown.

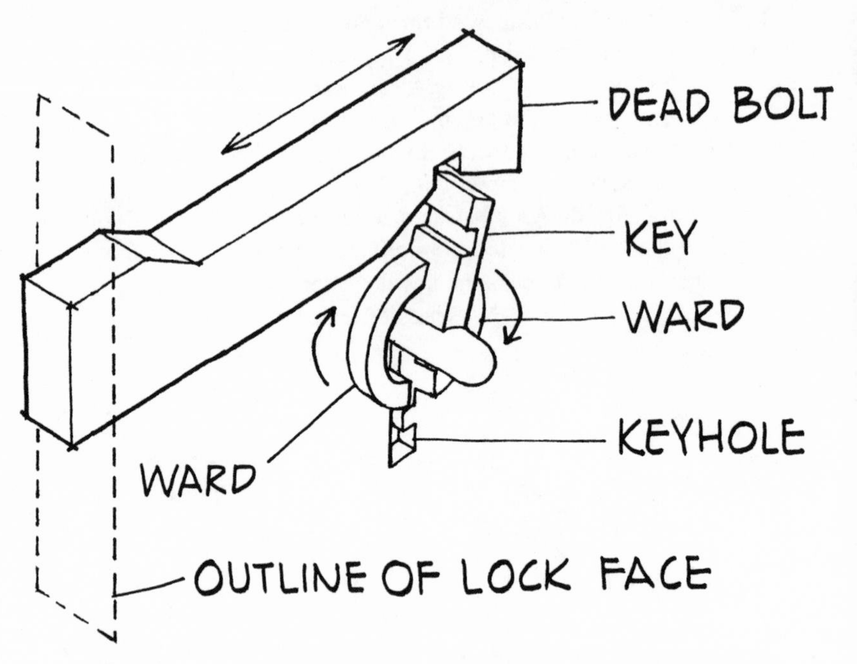

9. After the key has successfully entered the keyhole and has been turned past the side wards, the key's bit encounters the dead bolt. Turning the key to right or left advances or retracts the dead bolt, locking or unlocking the door.

INSTALLING A SURFACE-MOUNTED WARDED LOCK

An exterior-mounted warded lock is easily installed. The assembled lock is placed against the door at a height to place the center of the doorknob thirty-six inches above the bottom of the door. This is merely the conventional height today. The face of the lock (see Figure 10) is placed flush with the edge of the door, and a square is used to make certain the lock case is horizontal. Then a slim pencil or nail is used to locate the mounting holes. The lock is removed and pilot holes are drilled for the screws. The lock is replaced and temporarily fastened with screws. Now a mark is made for the spindle and another for the keyhole. The spindle hole is far

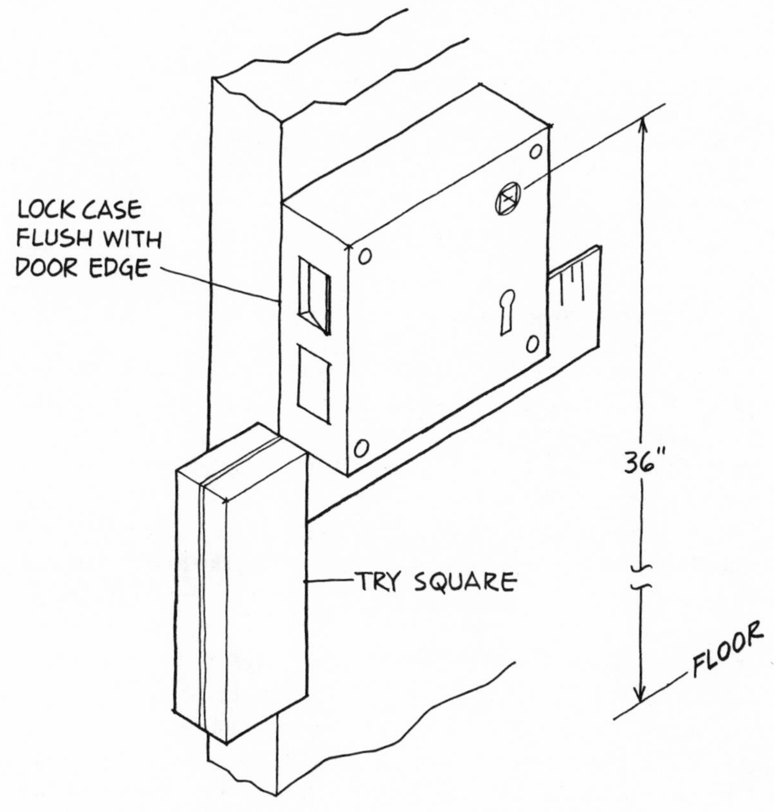

10. Using a try square to locate the lock on the door

from critical—generally a half-inch hole is plenty—but care must be exercised to make the hole for the key no larger than can be covered by the escutcheon.

The strike plate for the surface-mounted warded lock is held in place with wood screws and fastened atop the trim alongside the door jamb. To make the lock secure at this point it is necessary to use long screws. Therefore it is important that clearance holes be drilled through the trim. If you find there is no wood beneath that particular section of trim, you will have to remove the trim and fasten a strip of wood in place between the door jamb and the backing studs (see Chapter 8).

The strike plate is most easily located by closing the door and then extending the dead bolt as far as it will go. Now look between the door and jamb and mark the height of the latch bolt and dead bolt with a pencil. If possible, mark the inner edge of the bolts with a pencil. If not, darken the bolts after opening the door. Reclose, re-open, and note bolt marks. For best results the strike plate is mortised into the door frame. Clearance between the strike plate and the door should be held to under 1/16 inch if possible.

INSTALLING A MORTISED LOCK

Mortised lock installation requires more care and skill. The first step is to locate the spindle above the floor. Generally the spindle (and knob) center is placed thirty-six inches above the floor. A pencil line is drawn across the door at this height. Now the lock is placed flush with the door and its spindle hole is aligned with the line. Next its case is outlined with pencil on the side of the door, and then the top and bottom of its face are marked on the side of the door. The lock is put aside and the face marks and case marks are transferred to the edge of the door with a try square, then a pair of parallel vertical lines are scribed on the edge of the door. These outline the width of the lock case (see Figure 11).

The next task is to cut a rectangular hole in the edge of the door, just a little larger than the lock's case. Using a brace with a wood bit just a mite larger than the thickness of the case and a try square clamped to the side of the door to keep the bit orthogonal, drill two or more holes into the edge of the door. A wood chisel is used to remove the wood and cut a mortise for the face of the lock. Use a machinist's compass to transfer the backset and spacing distance for the spindle and keyhole from the case to the side of the door. Insert the lock into the mortise to determine the height of these holes.

To locate the strike plate for the mortised lock, which is also mortised in the door frame, mount the lock in its aperture and gently close the door. Use a pencil to mark the height of the dead bolt and latch bolt on the edge of the jamb. Then mark the closed edge of the door on the jamb, open the door, and transfer the distance from the side of the dead bolt to the side of the door to the door jamb. Draw the outlines of the bolts on the jamb, place the strike

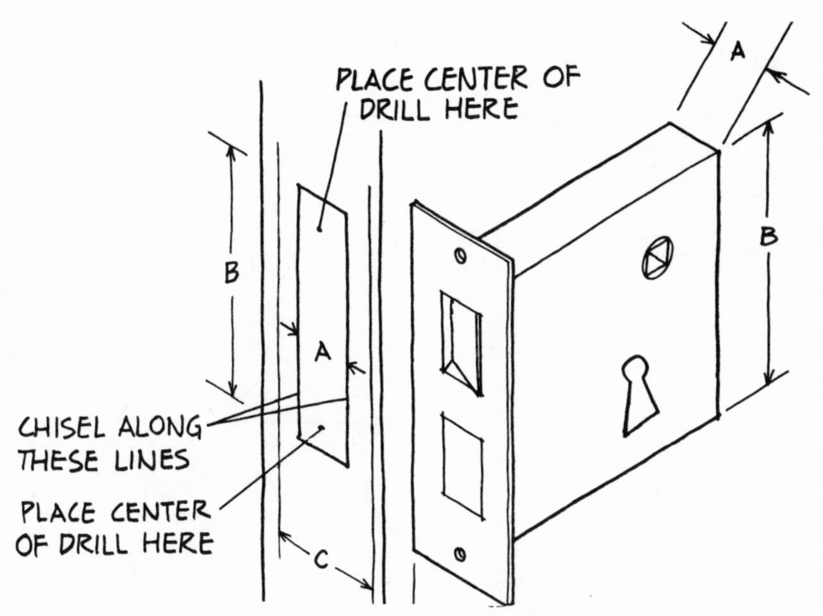

11. Installing a mortised lock. Place lock at correct height, its face flush with door. Carry outline across. Transfer A and B to door edge. Drill as indicated with bit diameter slightly greater than A. To mortise lock face, insert lock in hole, then lay out edges of necessary cuts.

plate over the outlines, and center it. Now you can outline the strike plate with a pencil. To be certain, you can darken the bolts with a pencil and then close the door, letting the bolts gently strike the wood. Correct if necessary.

A correctly positioned strike plate is shown in Figure 12. The strike-plate mortise should be cut to make the plate flush with the surface of the wood but no deeper. Use a drill and brace to cut the holes in the wood for the bolts. Take sufficient time with this mortise to make the strike plate snug. If it is not snug, the repeated opening and closing of the door will soon knock it loose.

ADJUSTING THE SPINDLE

Spindle adjustment is similar on both types of warded locks. Some old spindles have three screw holes on both ends of the spin-

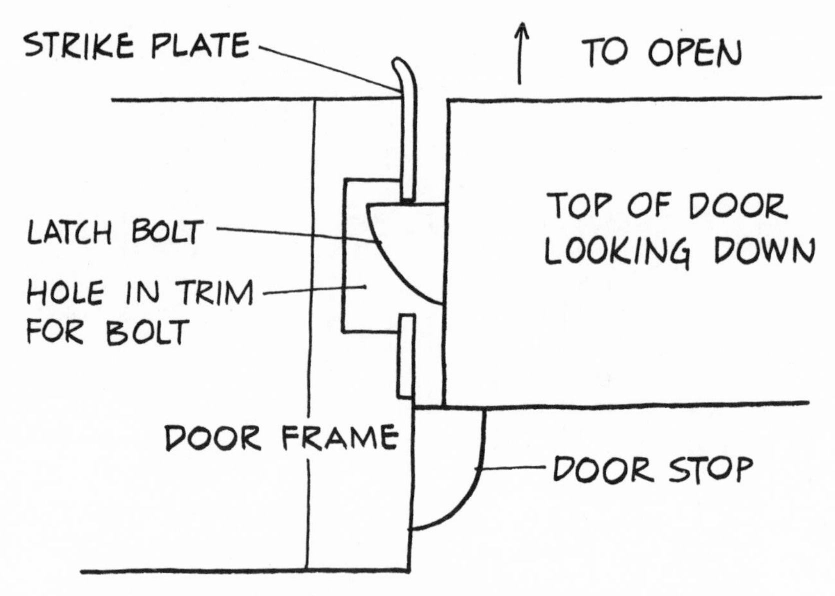

12. Top view of a correctly positioned strike plate. Note that while the hole in the strike plate is larger than the latch bolt, the plate is positioned so that the flat on the latch presses against the front inner edge of the bolt hole, and the rear of the door presses against the door stop. There should be no clearance in either of these two points; if there is, the door will rattle. If the strike plate has already been installed, loosen and move the door stop. On new construction the stop is not permanently nailed down until the strike plate is mounted.

dle bar. With these, one chooses the pair of holes which brings the knobs closest to the escutcheon without binding or flopping back and forth. The holes are not spaced equally apart so that a variation in knob separation distance can be achieved by selection. Other types have threads on both ends of the spindle. The knobs are screwed on until contact is made; then they are backed off a quarter-turn. Set screws on the collars of the knobs hold them fast to the spindle. There are also split-bolt types with two spindle sections, one screwing into the other to vary spindle length. First these spindles are adjusted for length, then the knobs are positioned. Some types have one knob permanently affixed to the spindle and only one knob is adjustable.

No matter what the spindle type, the knobs must be positioned close to the escutcheons but not so closely that they bind. If the knobs work loose, however, the spindle will slip to one side. With the split type of spindle, the lock may be operable from only one side of the door (see Figure 13).

Some of the warded locks are suitable for only one side of the door; that is, their spring latch bolts are not reversible. With other designs it is possible to rotate the spring bolt so that the door can be right- or left-handed. To ascertain this, remove the lock's cover carefully. Do not let any of the springs jump out; it may take some time to figure out how they return. Examine the spring bolt. If it can be removed and turned over on its other side, the lock can be used both ways.

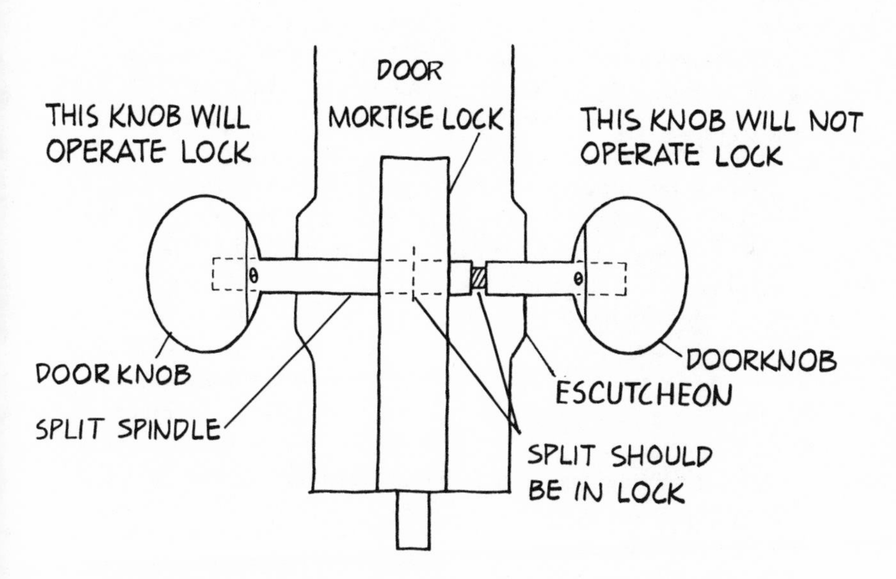

13. If the two halves of a spindle are not tightly fastened together, and the split is not in the center of the lock, as indicated above, one knob can be turned without operating the lock. Sometimes pushing the spindle to the other side makes the other knob operative. Remove one knob, adjust and tighten spindle, and replace knob. Leave some play between knob and escutcheon plate.

MAKING KEYS FOR WARDED LOCKS

Keys for warded locks are easily made and duplicated. The operation of the key as it is turned, moving the dead bolt, is so simple as to be self-explanatory. If the case cover can be removed, it is a simple matter to ascertain the necessary shape the key must have to get past the wards and reach the bolt mechanism.

If the lock's case cannot be removed, as would be the case if one could not or did not wish to remove the lock from the door, the first step is to select a key blank that will pass through the keyhole. The key is blackened by holding it over a match or over a candle for a moment. The key is returned to the lock and an attempt made to turn it. This must be done carefully: The key is not to touch the case while entering and leaving.

Where the key's bit has encountered a ward, a bright spot will appear. This is where the soot has been rubbed off. One then takes a needle file and removes some metal, then reblackens the key. Again the key is inserted without rubbing it against the lock's case or the keyhole. Again the key is turned and removed. Then the bright spot is carefully filed again. This is repeated again and again until the key passes the wards, can be turned, and opens the lock.

At first, the process will be found difficult and tedious. After some experience, keys for warded locks can be made in a short while with very few trials. Experienced locksmiths can make warded keys with little more than a glance at the keyhole.

Once one key has been made, duplicate keys can be made. This can be done with a bench vise and a suitable file. The only problem is keeping the two bits parallel. The solution lies in using a piece of aluminum or wood which is just as thick as the key's pin. Figure 14 illustrates the setup.

After the bitting has been cut in the key, the burrs should be removed to prevent them from catching on one's clothing or scratching.

TROUBLE-SHOOTING WARDED LOCKS

The difficulties most frequently encountered with door-mounted warded locks, as opposed to warded padlocks, are not due to the

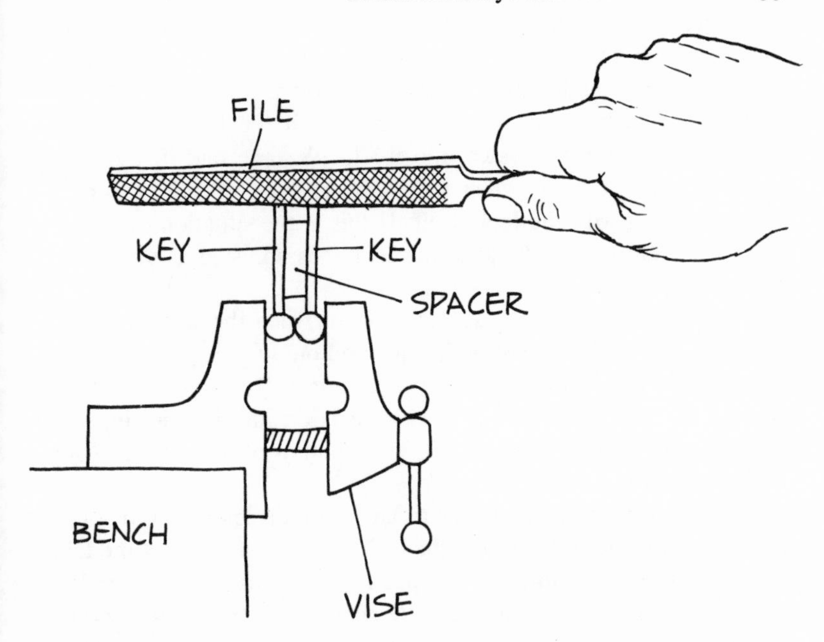

14. How a warded key blank may be cut to match an existing key. Wood or soft metal spacer keeps the two keys parallel while the new key is filed into shape.

lock itself but to its spindle and its knobs. For one reason or another, one or both knobs work loose with time, and they separate. Depending on the type of spindle involved, lateral movement of the spindle may disengage the square shaft from the square hole and the spindle can be turned without affecting the lock mechanism. Sometimes just one knob comes loose. Then the complaint is that the lock can only be operated from one side of the door.

Still another spindle-oriented difficulty is worn collars or escutcheons. In time these wear out, permitting the spindle to move laterally through lock and door. When this happens—depending on spindle design—the lock can only be worked when the knob is pulled or pushed in some mysterious way. This is when the smart-aleck kid gets his adult betters with the line, "See, I can make the door work. Why can't you?"

In any of the above instances the first step is the removal of one knob and then the spindle. A square-edged screwdriver is gently inserted into the spindle hole. The screwdriver is turned, and the action of the spring latch bolt—the bolt with the beveled end—is observed. If it can be drawn into the lock's case, and if it springs briskly back out again when the screwdriver is released, there is nothing wrong with the lock itself. If this is the situation, the spindle bolt is returned to the lock and its knobs adjusted as previously discussed.

If the latch bolt doesn't spring forward when the screwdriver is removed, the latch bolt spring has jumped out of place, has broken, or something has fallen into the mechanism. The lock has to be opened up. If the latch bolt movement is erratic, if it is sluggish, there may be paint on the bolt. Try scraping it off before removing the lock's cover.

With some models, the cover to an externally mounted lock can be removed without disturbing the lock proper, but in most models, it is necessary to remove the surface lock case to get at its underside cover. Generally it is a simple matter to remove the screws holding the lock to the door and then to remove the cover of the lock case; but in some instances the lock has been painted so many times it is difficult to find the screws, and even when you find them, they may have been so chewed up by previous efforts that they will not respond to a screwdriver. When this is the case, it is best to drill holes into the lengths of the screws and use a bolt puller, a special reverse-thread screw, to get them out. Tapping the edge of a screw with a prick punch works fine when the support is strong. Tapping an old cast-iron case may very well crack it.

If the screws are easily removed from a case but the cover is sealed in place with paint, apply paint remover to the joining edges. To restore the appearance of an old lock that has been painted to death, dip it into a can filled with paint remover and let it sit awhile. Afterward rub the paint off with a stiff brush, then remove the paint remover with alcohol or soap and hot water. Finally, dry and lubricate the lock.

To remove a mortised warded lock from its hiding place in the door, remove the spindle, key, and the two screws on the front of the lock. Using a knife point to avoid damaging the wood, gently pry the lock out. If time and hardened paint hold it fast, try cutting it free from the paint. If that doesn't do it, extend the dead bolt and catch hold of it with a pair of pliers and pull. If it is still stubborn, remove the escutcheons; there may be enough space between the spindle hole and the case to get a pry hold. Do not force a screwdriver or chisel behind the front of the lock. Some of the old locks have brass fronts, and once they are bent, it is impossible to remove the kink that is made.

Cracked cast-iron warded locks can be repaired by brazing, welding, or cementing. Brazing provides the strongest joint, but it requires that the entire piece of metal be brought up to the temperature of the brazing filler metal, usually brass. The problem here is that the cast iron may crack when this is done. Welding's heat is concentrated and therefore does not heat the entire casting as much. Most welders will tell you that cast iron cannot be welded, but this is not true. It requires a special technique. The weld should be made on both sides of the case, however, which requires difficult grinding or leaves an unsightly case.

Cementing with an epoxy cement is the best compromise, although the results are not always satisfactory from the point of view of strength. In theory, the epoxy is mixed and applied to the edges of the clean break. The parts are held in place with weights or clamps and then the assembly is placed in the oven, which is turned to 250 degrees or so. Depending on the type of epoxy used, setting may be completed in a few hours.

If the break is recent and the edges of the break are still bright and shiny, a good joint will usually result. If the edges have rusted or grease and oil from the lock have soaked into the cast iron, a strong joint is unlikely. In such situations it is best to back the joint up with a section of plastic or metal. The problem that arises in such cases is that there may not be room enough for the addition of a support. The solution may be to use aluminum and bend it to fit the weak section or to use a length of plastic and carve it to fit the area.

Whether you use a joint-supporting member or not, the metal case parts should be thoroughly washed down in strong soap and hot water and placed in the oven to dry thoroughly. Afterward the epoxy joint supported by the stiffener may be made.

Being ultra-simple, warded locks rarely suffer failure. When they do, it can most often be traced to paint dripping into the works, to rusting, or to broken springs.

As mentioned previously, paint can be removed with paint-remover solutions. Rust can be removed with modern removal solutions and sandpapering. When sandpapering, do not rub all the rust away. Unless you plan to lacquer the lock's works and exhibit them in your collection, stop sandpapering when most of the rust is gone and a little rust color remains. This oxide layer is most important to the life of the metal. If you rub it bright it will quickly oxide all over again. If you leave some rust you will have an absorbent surface which will hold a lubricant.

Any type of lubricant can be used, but professional locksmiths use special lock lubricants when they have them or Vaseline—petroleum jelly—when they don't. Oil is fine, but there is always the possibility it will drip out and soil the door or the floor.

Broken springs present a problem only in that they are difficult to replace. Find the broken pieces, if possible, and make a new spring from a length of clock spring. You can sever it by holding it firmly in a vise and forcing it into the sharp bend with a hammer. The nub that forms on the broken-spring end can be ground off.

Another problem resulting from a broken spring is part placement. Sometimes the parts fly apart when a spring breaks and it is a puzzle to fit them back together again. Since there are perhaps thousands of different warded lock designs, it is impractical to list all variations here. Just reassemble all the parts you are certain of, bearing in mind their operation. Then keep trying the remaining parts until you find their correct position and placement.

WARDED LOCK MAINTENANCE

To maintain a warded lock in good operating condition, the lock should be opened every year or so and dusted out. It is then given a touch of suitable grease and reclosed. At this time it is wise to make certain the strike plate is tight. Do not tighten the strike-plate screws or overtighten the screws when replacing the lock. Once a wood screw has been overtightened, it cannot be made tight again.

One cure is to remove the wood screw and dab some epoxy cement into the hole with a toothpick. Spread it around the inside of the hole. Use just enough to rebuild the wall threads, but not so much that the screw hole will be closed. Wait until the cement sets before replacing the screws.

Chapter 4

Lever Tumbler Locks

For some 1,600 years the warded lock was the only lock produced in the Western world. Smiths vied with one another in making their locks and keys more beautiful and their keyholes and warding more complex, but the lock mechanism remained the same. Then, sometime during the fifteenth century, a now unknown locksmith devised the lever tumbler lock. From a security point of view it was better than the warded lock, but not much. Meeting a lever tumbler lock for the first time, a picklock might find it difficult. On the second and third meetings it would present little if any more trouble than the warded lock.

The first lever tumbler locks had but one lever, and that lever had but a single action. To operate this lock the key needed only to lift the lever beyond a certain height. The single-lever design was popular almost down to the present day, and locks based on this principle are still to be found in old homes, bearing such famous lock-manufacturing names as Russwin and Schlage. Figure 15 illustrates a Russwin lock.

In 1778 the Englishman Robert Barron patented and began manufacturing his improved lever tumbler lock. The Barron lock differed from its predecessors in two remarkable and important aspects. It had more than one lever and the levers were double-acting. Each lever had to be lifted a specific distance. If lifted too much or too little, the key would not operate.

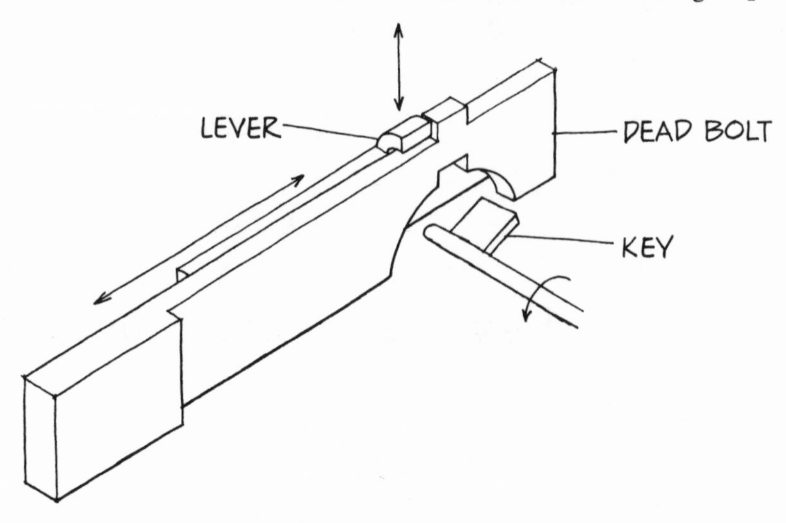

15. The workings of the Russwin single-lever tumbler lock. When the key is turned, the bit strikes the lever and lifts it. This frees the dead bolt, and the key is able to continue its turn and move the dead bolt, either into or out of the jamb.

From a security point of view the difference is tremendous. Assume ten levers—lever tumbler locks will have anywhere from four to ten levers—and assume that manufacturing tolerances permit ten different bitting heights. This too is a very realistic figure. Then the number of possible key combinations is 10^{10}, or ten billion. In practice the number is far less, probably on the order of thirty thousand to forty thousand, but still so many more than the warded lock and the single-lever lock that the Barron lock marks the beginning of the modern era of lock making.

For many years the public believed that the Barron lock could not be picked. Not until the American lock salesman Alfred Hobbs came to England in 1851 and applied his "tickling" or "tentative" method to the Barron lock was it publicly picked.

Figure 16 is a sketch of what is reported to be Barron's original lock. The basic design is still used today. Today's locks, however, have their levers mounted on the dead bolt. Barron supported his two levers by a shaft fastened to the lock's case. This is not to imply that our inventors quit after Barron created his masterpiece—far

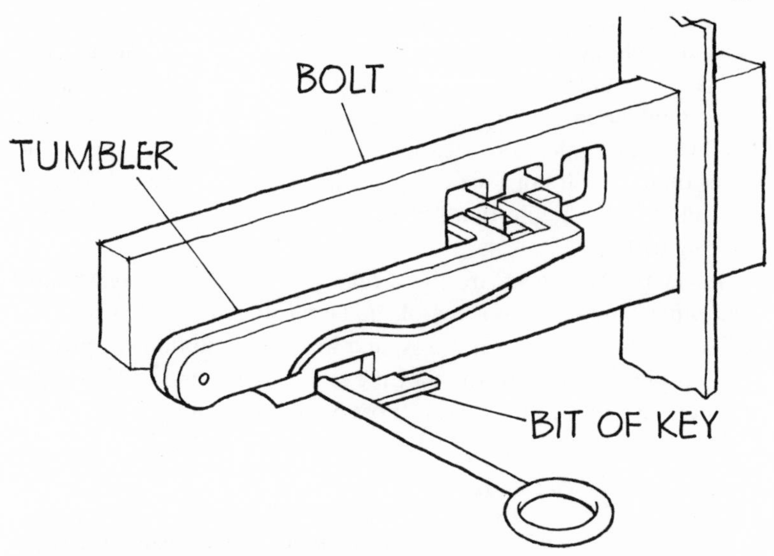

16. Barron's 1778 lever tumbler lock (after a drawing by F. P. Gillman). There are only two levers.

from it—but subsequent lever tumbler inventions improved on the basic design by adding components; none altered the Barron concept.

Hobbs would not have been able to use his method to pick a Barron lock had it been equipped with false gatings or notches invented by a man named Russell (who worked for Bramah). Hobbs himself invented a "safety lever" that rendered his "tickling" method inoperative, and Jeremiah Chubb varied the Barron lock with a "detector." The detector comprised a "latch" that caught and held any lever raised too high by either a picklock or the wrong key. Thus the boss, arriving late at the office, could tell whether or not any of the help had been at his safe and, with some models, how many times.

The better, more secure lever tumbler locks are large and appear to be far sturdier than the smaller pin tumbler locks, which are discussed in Chapter 6. Lever tumblers are not as easily clogged with chewing gum or kneaded bread, and duplicate keys for lever tumbler locks are not as easily secured as are duplicate keys for pin tumbler locks. It is not that a lever tumbler key is more difficult to duplicate, it is simply that far fewer locksmiths are prepared to duplicate such keys because they rarely stock lever tumbler key blanks. Still another possible reason for greater security is that a lever tumbler lock is not as easily picked by a beginner as a pin tumbler lock, which, on occasion, will open to a sharp "rap." The lever tumbler will not (see Chapter 10). Experts believe, however, that ultimately the pin tumbler has greater security and withstands picking better.

Lever tumbler locks are to be found in the doors of old homes and other buildings, on some of the better attaché cases, suitcases, cabinets, and safe deposit boxes, and in such institutions as prisons and hospitals, although why they are still installed and used in these institutions is not entirely clear. Possibly it is custom.

PRINCIPLE OF OPERATION

Figures 17 and 18 illustrate the workings of two general designs of present-day lever tumbler locks. In the first figure the lever and the post are both fitted with notches. This reduces the possibility of the gate slipping past the edge of the post and makes it more difficult for a picklock or the wrong key to operate the lock.

In the first figure only one lever is visible, but there may be any number of levers. The more levers, the better the lock. Commonly any number from four to six may be found with a lock of this type.

Figure 18 illustrates another Barron variation. The gate is an H-shaped hole in the levers. Dead bolt movement is possible only when all the levers are lined up with the center slot in the gate. When the bolt is home and the door is locked and the key is removed, the levers fall down. The dead bolt cannot be moved. When the bolt is withdrawn, the door opened, and the key removed, the levers fall down on the other side of the post. In this way it is impossible for someone to shake the door and jar the bolt halfway in. Should this happen, the key would be unable either to open or to close the bolt.

The same design produces another condition, desirable for safe deposit boxes and the like. The key cannot be withdrawn when the bolt is in the withdrawn position. In other words, the holder of the safe deposit box cannot return his box to the bank-vault wall and

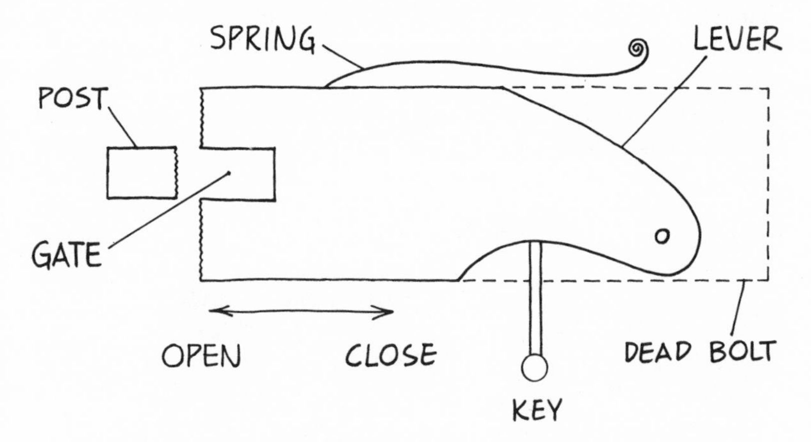

17. Only one lever is shown in this view, but there can be as many as ten more behind this one. Note how the key must lift the lever to the proper height to permit the post to enter the gate. Note serrations on gate and post, making a sloppy fit more difficult and thus lessening the lock's susceptibility to picking.

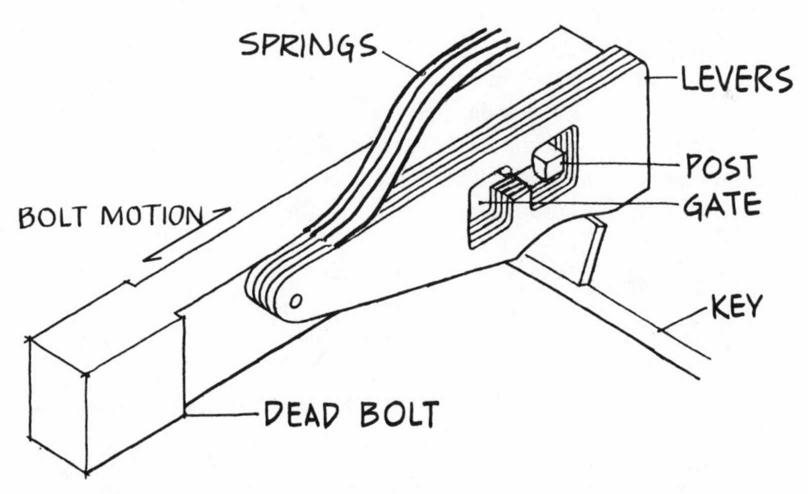

18. Double-acting multiple-lever tumbler lock. Note the relation of the gate to the post, and the spaces provided to either side of the gate into which the post may fit when the levers drop in response to the spring.

remove his key without locking his box. Should he forget, the key would remain in the box to remind him. In many banks the vault attendant accompanies the box holder into the vault. The box can only be unlocked with two keys used simultaneously. The attendant keeps one key.

INSTALLING LEVER TUMBLER LOCKS

Like warded locks, lever tumbler locks designed for doors are either surface-mounted or mortised. The overall shape of these locks is much like the overall shape of similar warded locks. In fact, it is difficult to differentiate between a warded lock and a lever tumbler lock at first glance. The directions given for installing surface-mounted and mortised warded locks in Chapter 3 apply here.

DUPLICATING LEVER TUMBLER KEYS

The method used for duplicating lever tumbler keys by hand is simple enough. The work, however, is tedious and exacting. The key and the blank are placed side by side in a vise. A narrow file is used to cut the bitting on the blank. Great care must be used, because the width of the cut as well as the depth of the cut is critical. If the cut is made too wide, two levers may enter it, possibly jamming and possibly becoming bent when the key is pressured to turn. If the slot is cut too narrow the lever may not enter.

If possible, lever tumbler key blanks should be secured from the manufacturer or from a local locksmith. In the latter case it may be advisable to have him cut the new key on his machine. On the other hand, if a blank is not available, an old large key of the same thickness may be cut to fit, or a sheet of metal may be used. The metal must of course be of the correct thickness and preferably of a metal that is "usefully" hard. If the metal is too soft the key will wear out very quickly. If it is too hard it will be difficult to cut and it will wear the levers. This is not too important if there are no other keys to the lock, but if there are and the levers change appreciably, the other keys may not fit after a period of time. Mild steel, iron, bronze, brass, and high-temper aluminum are suitable key metals.

FITTING KEYS TO LEVER LOCKS

In order to fit a key to a lever tumbler lock it is necessary to see its innards. This is easily accomplished with locks fitted with windows for key-making purposes. The lock is removed from the door and the window cover is removed. When the lock does not have a window, the case must be opened. While this can almost always be done (the exceptions are the riveted locks), it is inadvisable to open a lever lock without specific need. Since the case cover holds parts in place, springs may fly out and be difficult to replace.

Using a length of stiff wire or a slim screwdriver, the levers are gently manipulated one by one until the dead bolt is in the locked —extended—position. This is more easily done if the lock is pointed downward, its bolt toward the ground.

A suitable blank is secured or prepared. It is blackened by placing in a candle flame or a match flame. Then it is inserted into the lock and given a slight turn. All the levers will indicate their presence and position by removing some of the soot where they contact the key blank. At the moment we are only interested in the lever nearest the window.

The key blank is removed and sufficiently filed down under lever number one, the nearest one, to permit that lever to drop to where its gate is in line with the post. This is best not done at one time but by repeated filings and examination. Then the blank is cut for the second lever.

Great care is exercised in filing the blank to make certain that the cuts are directly beneath their respective levers and that no cut is so wide as to accept two levers, or so narrow as to catch a lever and stick. It is best to use a file that is thinner than the lever.

When turning the blank to check on the height of the levers, turn slowly. If the cut is crooked—partially out of line—the lever may be bent. This will necessitate dismantling the lock and straightening the lever. When the lever is thin and there is little space between it and its fellows, even small kinks in the metal can be troublesome.

FITTING LEVER TUMBLER LOCKS TO KEYS

The alternative to cutting a key to fit a particular lock is cutting a lock's levers to fit an individual key. This can be done to any lever lock whose levers are accessible and removable, and it can be done to any set or group of levers that have not already been too deeply cut to accommodate the new key.

The first step is to make the gates and post visible. The second is to try the new key, turning it gently as far as it will go and carefully observing the gates. If the first cut on the key, the cut nearest the bow, is deep and the gate on the first lever is not raised to or above the post, that cut on the key cannot be accommodated by that lever. Further lever cutting will merely lower the gate. If extra levers fitting that lock are available, the first lever can be replaced. If no additional levers are to be had, inspect the balance of the gates. If one gate is very high, it is possible that it can be interchanged with the first one.

Assuming that extra levers have been used as needed or that changing lever positions satisfied the requirements, what if you have a group of levers that are all too high, their gates above the post? The height of these levers must be reduced by filing. Two words of caution. First, the position of the key's bitting in relation to the levers and their pivot points usually makes for a 2 to 1 or even a 3 to 1 ratio between the metal removed from the lever at point of key-to-lever contact and the position of the gate to the post. In other words, a little filing on the lever may make considerable difference in the position of the gate. Second, the key must slide across the edge of the levers. Filing has to be smooth and even, with no sudden changes in lever height.

As each lever is filed, it is returned to its correct position in the lock and tested with the key. It is best at this time to return the lever springs to their positions. After all the levers have been changed, the lock is reassembled.

LEVER TUMBLER LOCK REPAIRS

Broken and disconnected lever springs are the most frequent cause of trouble with lever locks. Each lever has a spring. Should the spring break or jump out of place, the lever may stick in an up position, in which case the lock will be inoperable. If a piece of spring gets between the dead bolt and the case, the lock may be difficult to open and close. If it gets very bad the key may be bent. Although all the lever springs cannot be seen from the window in the lock case, their action can be judged by watching the key work the levers. This action should be smooth and uniform. Another test is to remove the key and tilt the lock's case. The lever without the spring will fall back when the case is tilted. If it is necessary to remove the lock's case to inspect its insides, be careful when you tilt the case, for parts may fall out.

Lever lock springs are usually made of fine flat spring steel. They jump out of place because they have lost the tension they need to remain in place; they were too short to start with, they were not properly shaped, they have been kinked or improperly placed or bent by a previous mechanic, or they have broken. Springs break because of excessive grain boundary dislocations. A tiny crack caused by impurities in the metal or by improper tempering propagates until the metal fails. Metals do not break because they have crystallized. In their solid states all metals are normally crystalline.

Take care when removing levers to keep all the levers in their proper order. If the order is changed, the original keys will not operate the lock. Note that levers are often marked with numbers, but these numbers do *not* refer to the position of the levers in the lock. Instead they indicate the relative key cut depths. A shallow cut would be designated by "1," a very deep cut by "9," and other cut depths by numbers in between.

Replacement springs may be cut from piano wire. They should be cut to length and bent to approximate shape before insertion into the lock. This is done to avoid placing any strain on the levers, which could bend them. Choose a wire thickness that will provide spring pressure equal to that originally used. A reasonable change in pressure makes little difference, but too much pressure will place an undue load on the key and shorten its life. Overly high pressure will not necessarily insure the spring end's presence in its slot. It may tear itself loose. On the other hand, too little pressure will fail to operate the levers properly.

The end of the spring that makes contact with the lever fits into a slot in the lever. Make certain this slot is clean and that the new spring is thin enough to fit properly. On some locks the edges of the slot are rolled slightly up and over the top surface of the spring end. This is done on a clean flat metal surface with a small hammer.

Care is needed to make certain the lever is not mauled and bent out of shape and that too much metal is not rolled over the spring—so much that there is nothing but a thin shell holding the spring. After peening, the spring and lever are returned to the lock together.

Worn levers are another source of trouble. The key has in time worn so much metal from the lever-contact surfaces that the key no longer raises the gates to the height of the post. The first indication of this is that the key becomes increasingly more difficult to turn. A look through the lock's window will quickly ascertain the difficulty. If it is wear, all the gates will be low.

If the key or keys that operate the lock with the worn levers do not have to operate other locks, as they might if they were master keys, the simplest and perhaps the best way to remedy the trouble is to cut a new key. The bitting on the new key is cut slightly higher to offset lever wear. If this is not possible, the levers are removed and replaced, or the levers are placed on a clean, flat metal surface and lightly pounded at the wear area. This makes the levers wider but thinner. Next a file is used to remove any rough spots on the working edges of the levers. Care must be taken not to deform the metal when pounding.

A poor alternative sometimes used by second-rate mechanics is widening the gate opening. Doing so makes the lock responsive to more keys and greatly reduces its security. Sometimes one will find locks from which levers have been removed, again as a rapid means of curing an ill. Reducing the number of levers in a lock reduces security directly, although in some instances it may be advisable to remove levers. For example, there is no need for a six-lever lock to be installed on a cupboard. If that is the only lock available, some levers could be removed to simplify key making and master keying.

Occasionally, lever tumbler locks will be found with worn lever pivots and worn pivot holes in the levers. Whether or not the increased play is important depends on the relation of the lever's spring and the pivot hole and the key's bitting. If the spring rides on the lever at a point between its support and the highest point of the key's swing, the increase in hole clearance will have no more effect than that of a worn key or a worn edge on the lever. If the spring's pressure point is on the gate side of the key's contact area, the lever will tilt from side to side as the key swings under the spring's end. When this happens, a slight amount of pivot-hole wear results in a considerable difference in gate positioning. The simplest repair is to alter the spring to bring its end closer to the pivot point. If this is not possible, pounding the hole area on the lever will decrease the diameter of the hole. A new lever is the best remedy, followed by electroplating. There are many electroplating shops that will apply a layer of nickel for a reasonable fee. This may be easier than cutting a new lever from a sheet of hard brass or other metal.

MAINTENANCE OF LEVER TUMBLER LOCKS

Lever tumbler locks require no more than a puff or two of graphite every year or so. If some of the parts are of iron, these parts may be wiped down with light oil and then wiped dry after a few hours. The oil will soak into the light rust and prevent—to some degree—further rusting. A drop or two of fine oil may be used in and around the bolt mechanism, but no oil should be used on the levers (unless they are of iron, and then they should be wiped dry) or their bearings. Generally there isn't too much clearance here, and the best oil dries with time. Graphite, on the other hand, does not deteriorate with age and does not harden and thicken.

Chapter 5

Disk Tumbler Locks

The disk tumbler lock was developed fairly recently in answer to the demand for a small low-priced lock that has a reasonable degree of security, which we have seen the warded lock does not.

The disk tumbler lock is easily installed and removed. It may be found on almost all automobiles, many metal office desks and filing cabinets, slot machines, vending machines, and similar devices. It can be recognized by its key, which while similar to the pin tumbler key is usually far shorter in length but equally broad. A disk tumbler key may be bitted on one side or both. Disk tumbler keys usually have very simple side ward indentations. The keyways usually look like the letter "i."

The best of the disk tumbler locks are less than half as secure as the best of the pin tumbler locks. And since disk tumbler locks are rarely if ever produced without the designer's eye on the selling price, no disk tumbler lock, to this writer's knowledge, has been constructed that exhibits the lock type's maximum security. It therefore would not be far wrong to ascribe to conventional high-quality disk tumbler locks no more than one-tenth the security possible with high-quality pin tumbler locks. In numbers this means the disk lock is capable of differentiating between no more than three thousand to four thousand different keys and possibly fewer. The pin tumbler can differentiate between thirty thousand and forty thousand different keys.

PRINCIPLE OF OPERATION

In principle of operation the disk lock is similar to the pin tumbler lock, which is discussed in the following chapter. Imagine a number of disks placed face to face. Each disk is pierced by a rectangular hole. Each hole is identical in width but differs in height. Each disk has a projection on top, which also differs from disk to disk. Now we take a slim metal bar and run it through all the holes in all the disks (see Figure 19). All the disks hang on the bar. Because of the height of their holes, they hang differently; their tops are at different heights. Since the projections on each disk are of different lengths or heights, the topmost edges of the disks are not flush. Holding our disks on the bar, we place a second bar above and to the left of the disk projections. If we try to turn the first bar with the disks to the left, their projections encounter the second bar. Now let us remove the first bar and replace it with a strip of metal that has been cut (see Figure 20). When we hang the disks on this "cut" bar, we find that the projections are now in a line and that this line is below the position occupied originally by the second bar.

Call the "cut" bar a key. Insert the disks in a metal shell. Position them with springs so that they are forced to move upward. Insert the metal shell into a second shell having a longitudinal slot in its interior. Each disk projection now engages this slot. The inner shell cannot be turned. When a properly cut length of metal ordinarily called a key is inserted with its cuts or indentations pointing down, it forces the disks down just enough to free them from the slot. The interior shell can now be turned. When it is turned it moves the dead bolt.

Figure 21 shows the interior of a disk tumbler lock with eleven disks in the locked position. A sort of safety pin pushes the disks outward against the inside of the case or shell. Figure 22 show what happens when the proper key is inserted: The disks line up; none stick up or down. Figure 23 is another lock of this type. Note how the disk projections protrude beyond the inner shell. To the right the disks themselves are shown along with a single disk. The spring catches in the side projections.

The inner shell, or plug, is held in place by any number of different means. In some designs a crescent spring is used, in others a flat "E" ring is used, and still others employ a screw threaded into the

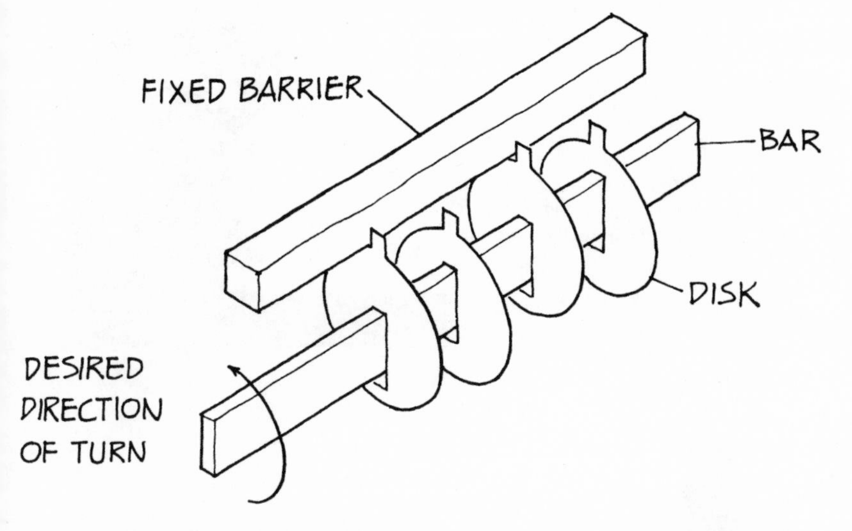

19. Principle of the disk lock. Teats on top of disks 1 and 3 project so high that they strike fixed barrier when an effort is made to turn supporting bar.

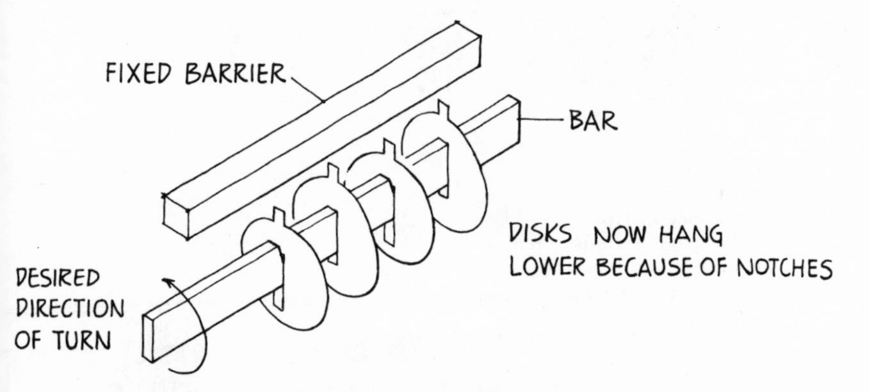

20. Same disks. Supporting bar, now called key, has been notched beneath disks 1 and 3. Tops of all disks are now below barrier. Supporting bar, or key, can be turned to left as desired.

21. Interior of a disk tumbler lock with eleven disks in locked position

22. Same lock with proper key inserted. Note how all disks line up.

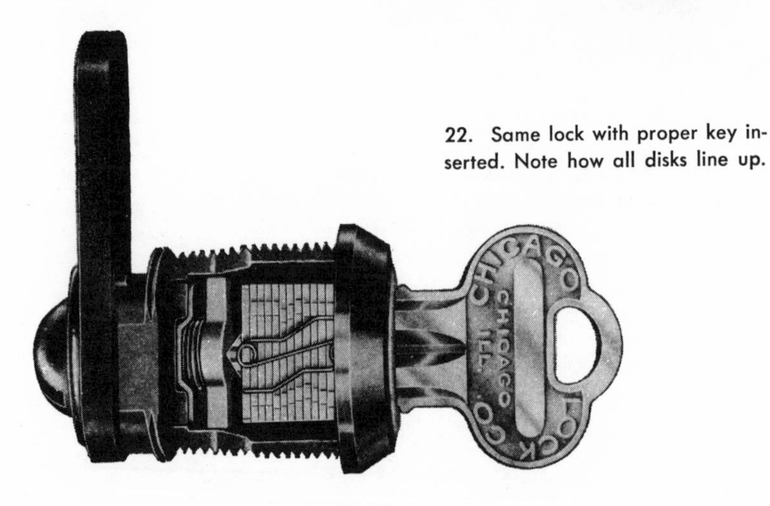

23. The same general type of lock with the plug removed from its shell. Note shape of individual disk to right. Springs fit between projections.

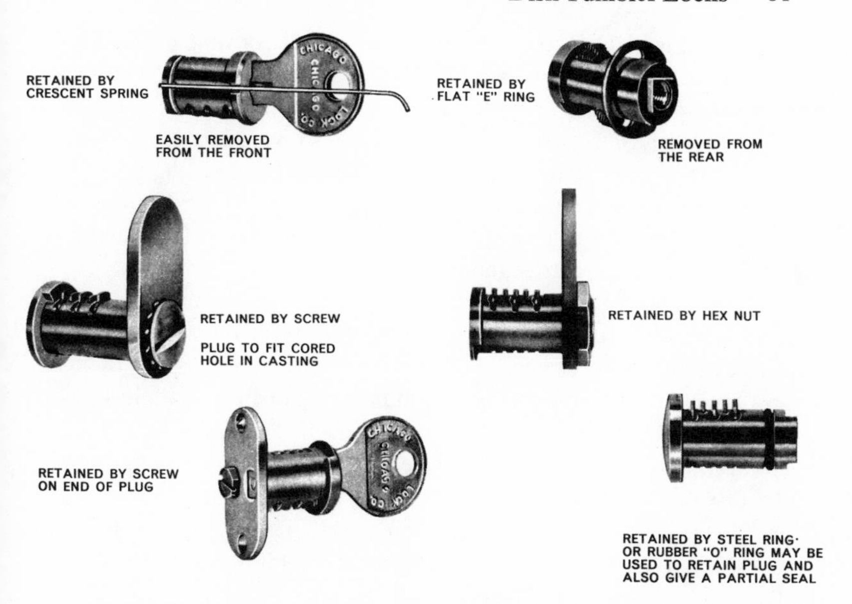

24. Various methods and devices employed to retain plugs in their housings. Plug at top left can be removed from the front when one has the correct key and tool. *(Courtesy Chicago Lock Co.)*

end of the plug. Some of these are illustrated in Figure 24. The plugs held by springs can often be removed without dismantling the lock itself. The proper key is required and it must be in place; then the release device is inserted and turned.

DUPLICATING KEYS FOR DISK TUMBLER LOCKS

Disk lock keys may be cut from suitable blanks by hand. The blank is aligned with the original key as carefully as possible and both are placed in a vise. A fine file is used to cut the second key until it matches the first.

CUTTING A KEY TO FIT A DISK LOCK

The plug is removed from the lock and placed in a suitable vise. The correct blank, secured from the lock's manufacturer or a local locksmith, is blackened by insertion in a candle or match flame for a moment. The blank is carefully inserted into the lock. The distance the first disk is extended by the presence of the key is noted. The easy way is to take this distance off with a machinist's compass. The key is removed and placed in a small vise. The point of contact between the inside edge of the first disk and the key will be visible in the form of a bright spot. If it is not, repeat the blackening and insert and remove the blank again.

Next a fine file is used to cut a "V" groove across the bit of the key blank. The bottom of the groove should be slightly rounded, as it will no doubt be if a fine jeweler's file is used. The depth of the groove should be something less than that indicated by the compass. This is a precaution. The key is inserted and the first disk's new position noted. If it still projects beyond its operational line, the first cut has to be deepened. If not, the key is again darkened and the second disk's position noted and cut. This is repeated for all disks. Then burrs are removed from the key's bitting, and the job is done.

FITTING THE DISK LOCK TO A KEY

A disk lock may be altered to match a new key that doesn't operate that lock. One method will be detailed here, but it should be noted that the cost of these locks is so low that generally what you save by changing a lock to suit a key is not worth the effort. However, there may be times when one wishes to make a number of locks respond to one key, and then unless a set of such locks is purchased, the locks will need to be altered.

The plug is removed from its shell and placed in a suitable vise. The new key is inserted. Note is taken of which disks are now too high or low (assuming it is a double-bitted lock). "Low" means that the disk projects too far below the plug wall. The disk is marked by counting its position from the rear of the plug. The retaining spring is removed and the plug disassembled. It is important that the disks be maintained in their correct order. The desired disk is removed and placed in a vise. Next the interior disk's edge, top or bottom, is filed the requisite amount. This will be the side of the interior rectangle opposite the overly high projection. The disks are reassembled and the new key tried. This is repeated until the presence of the new key brings that particular disk into line with the other disks.

It is unwise simply to file the offending disk projection down, because that disk will no longer function and the lock's security is reduced.

DISK LOCK REPAIRS

The most frequent cause of trouble in the disk lock, as in the lever tumbler lock, is its springs. These break and jump out of place. Sometimes they wedge between the disks and the inside of the cylinder, and when the plug is forced to turn, the disks or the key is damaged.

Broken springs must of course be replaced. Springs that have moved out of position must be replaced, but some effort and time should be spent to determine the cause of spring movement. There is no point in replacing a spring only to have it jump free after a few days. Examine the other springs to see how they are held in place. It may well be that the disk to which the loose spring was attached is broken or so badly worn that it slips from side to side.

Bent disks should not be returned to the lock but should be replaced. Proper lock operation permits little or no clearance between disks. If one or more are bent, they will stick, making key operation difficult.

DISK LOCK MAINTENANCE

Nothing need be done to these locks except lubricate them every few years or so with a few puffs of graphite. Oil should not be used. When operation becomes stiff, try graphite. If that doesn't help, check the key. It may be so worn that force is necessary to operate the lock.

Chapter 6

Pin Tumbler Locks

The pin tumbler lock is the most popular type of lock mechanism in use today. It may be found on doors to homes, offices, public buildings, and cabinets. It is used wherever a lock is desired with something less than the near-ultimate security of a combination lock or the ultimate protection of a time lock. Most practicing locksmiths believe that no other lock design approaches the pin tumbler in regard to cost, size, simplicity, service life, and small key.

No one to this writer's knowledge has undertaken a service-life study of pin tumbler locks, but from experience we know that even the low-priced exterior door locks of this type will last ten years or more. Some of the better pin tumbler locks, especially those with steel balls under their pins, have outlasted their owners and are still in operation.

People who live in apartment houses and find they can get through the entrance-hall door with a nail file may doubt this statement. These door locks are master-keyed, as apartment-house entrance doors must be, and are therefore comparatively short-lived. (Master keying and some of its attendant problems are discussed in Chapter 9.)

One wonders why the pin tumbler disappeared from the scene for so many generations. It is an exceedingly clever but simple machine —so simple that it can be fully appreciated only when one completely understands the way it works.

PRINCIPLE OF OPERATION

Figure 25 shows a number of keyholes typical of this type of lock. Figure 26 is a cross-sectional view of the basic pin tumbler mechanism. The perspective of the drawing has been distorted somewhat as an aid to clarity. Space is shown between the various parts. In actuality, these parts are close-fitting, with just enough clearance to move easily.

In this illustration note that there are five sets of pins, drivers, and springs and that they are held in place by a cover plate. All pin tumbler locks do not have a cover plate, which offers an easy and convenient means of removing and replacing the parts in the pin holes. Also note the core, or plug, the inserted key, the shear line, and the cam, or tail, which in normal operation is held by machine screws to the end of the plug.

It is this cam, or tail, that makes contact with the dead bolt. When the correct key is turned, the cam turns with it and moves the bolt forward and back, in and out of the door jamb, locking and unlocking the door.

The key shown in Figure 26 has been cut from the correct blank. It belongs to that lock's family of keys; that is, it fits the lock's keyway. This particular key will not operate the cylinder shown, however. If you examine the pins and drivers you will see why.

Reading from left to right, the juncture between pin 1 and driver 1 is on the shear line—the space between the plug and the shell. Pin 1 and driver 1 will not stop the plug from rotating in response to key movement.

Consider pin 2 and driver 2. Note that the shape of the key—the cut on its bit at that point—permits the pin and driver to move down to where the driver crosses the shear line. The plug cannot turn because driver number 2 is across the shear line, locking the plug to the shell. Further examination will reveal that drivers 4 and 5 have also been depressed by their associated springs to where they too cross the shear line and enter the plug.

To force this particular plug to turn in its cylinder with the key shown, it will be necessary to "shear" four steel pins. Each pin is about $\frac{3}{32}$ inch in diameter. No brass key will stand up to the force

Manufacturer
BARROWS
BEST
BRIGGS & STRATTON
CORBIN
EAGLE
KEIL
LOCKWOOD
MASTER
NORWALK
PENN
READING KNOB (WKS)
RUSSWIN
SAGER
SARGENT
SCHLAGE
YALE

25. A few examples of the many keyway cross-sections used with pin tumbler locks. None of the key blanks fitting any one of the above keyways is interchangeable.

necessary to do this. If a steel bar is driven into the keyway and forced with a wrench, the brass plug will give way somewhere along its length before all the pins are sheared. Most likely the pins will bend somewhat and jam between the deformed core and the deformed pin block. This is the source of the pin tumbler's tremendous strength.

In Figure 26 the key shown is correct for the accompanying pin tumbler setup at only one point, pin 1. This particular key has been cut too low at the other four pin locations. Let us examine the lock in Figure 27.

Again the wrong key has been inserted into the cylinder. Can you see why? The key has the correct cross-section, and it is neither too long nor too short. If it were too short there would not be a section of key under the first pin tumbler. If it were too long, some of the key's bitting would extend beyond the core of the lock. The reason this key cannot work can be seen by following the shear line. Note that pin 1 is above the shear line. Pin 2 and driver 2 are in line, meaning that the cut in the key's bit is correct at this point. However, pins 3, 4, and 5 are above the shear line.

The point behind the second lock-interior illustration is to show that the pins, the lower lengths of metal in the pin holes, can also prevent the plug from rotating.

When the pin and its driver do not meet at the shear line—when they meet above or below this line—rotation cannot be accomplished. In a five-pin cylinder the five pin-driver pairs must be at the correct height to allow plug rotation. In a four-pin-driver cylinder the four must be at the proper height for the lock to work.

Figure 28 shows the same lock with the correct key in place. Note that the drivers and pins meet each other at a common line and that this line is the shear line—the separation between plug and cylinder.

The drivers in the foregoing three illustrations are not of the same height. They can be, but often their lengths are varied to suit the pins they drive: a long driver for a short pin and vice versa.

Most often the tops of the pins are square with their lengths; however, some manufacturers round the tops of their pins to conform with the curvature of the core. Since the edges of the pins will wear down to eventually conform to this curve, the thought is that

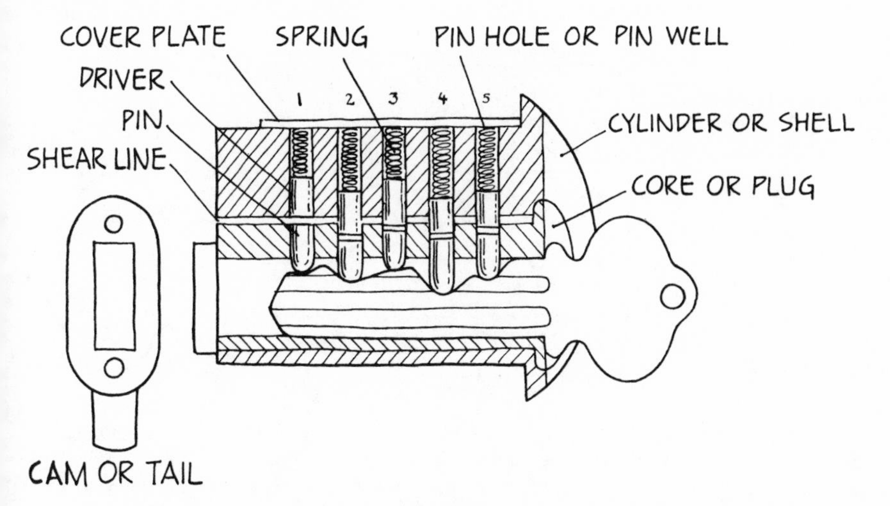

26. Cross-section of a modern pin tumbler lock. This lock has a cover plate which can be removed to expose the springs. There are five pin tumblers and therefore there may be five cuts on the key. The key shown will not operate this cylinder.

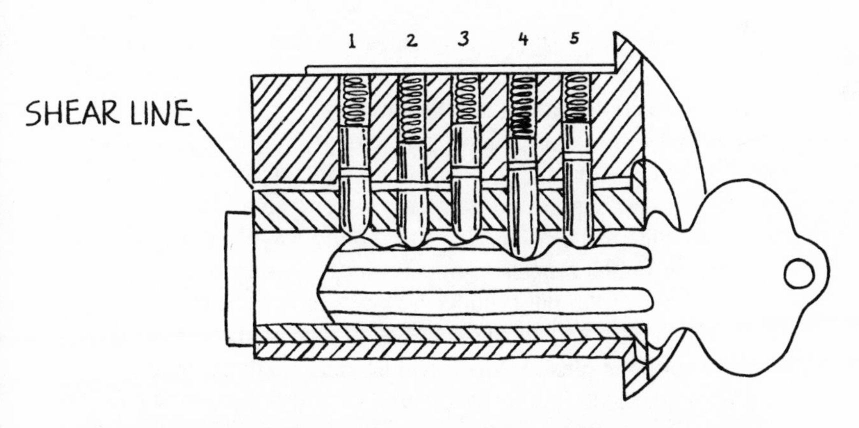

27. This key will not operate this lock. Note that tumbler pin number 1 is up in the shell. Number 2 meets the driver at the shear line. Pins 3, 4, and 5 are also well up and into the shell.

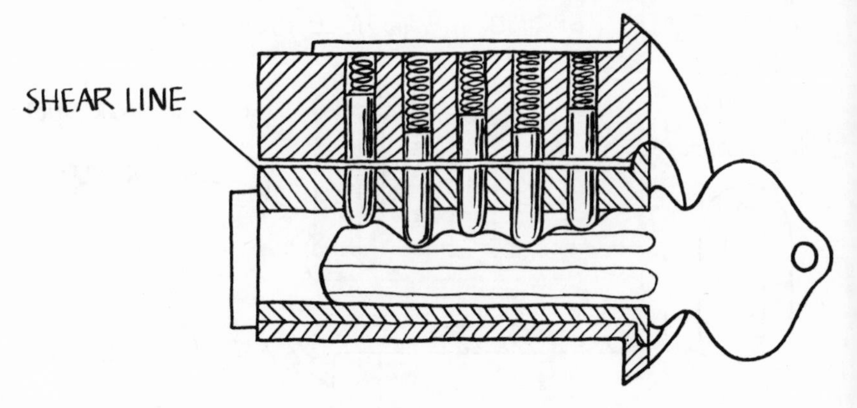

28. The key shown will operate this lock. Note that all the pins meet at the shear line. Normally the drive (top pin) contacts the tumbler pin, but they are shown a distance apart for clarity. Note that the longest tumbler pins are associated with the shortest drivers.

precurving will reduce the amount of change that occurs in the lifetime of the lock.

Other manufacturers use steel ball bearings to reduce the wear on the key and the pins. These bearings are placed under the pins, one to a pin, and they move up and down as the key is entered and removed from the lock. The balls do not change the operation of the lock, but they cannot be removed without a corresponding change in the key. Generally, the bearing nearest the front of the lock can be seen by looking into the keyway.

The springs provide positive pin and driver movement. Springs make it possible for pin tumbler locks to operate in any position. In the illustration the pins and drivers are shown in a vertical position, but in practice the lock may be turned any way desired. In a padlock the pin tumbler mechanism is most often sideways.

In the case of the pin tumbler lock, we have seen that to be correct the key must (1) fit the keyway, (2) be of the correct length, and (3) raise the pins to the correct height—no more, no less.

Pin tumbler locks are manufactured with as few as three pins and as many as eight pins; so-called good locks have five or six pins. Normal manufacturing tolerances permit notches, or key bitting, to be any of ten different heights—from zero to nine different changes. Assuming our lock has six pins, there are in theory 10^6, or one million, different keys possible with just one lock. Since there are some twenty important lock manufacturers in this country alone, each producing pin tumbler locks with no fewer than five differently shaped keyways, the total possible number of key changes (different keys and associate locks) rises to an astronomical number—$5 \times 20 \times 10^6$, or one hundred million. Lest this space-age number make you feel over-secure, we would like to point out that many professional locksmiths believe that the maximum number of practical key changes possible is in the neighborhood of forty thousand, which is still considerable.

The reasons for the great difference between theory and practice are these. There cannot be too great a height difference between any two neighboring pin cuts or the pins will "hang" up at that point. In other words, you cannot make the first cut 0 and the following cut 9. And you cannot make the last cut, the one nearest the bow, a 9, because it will be too thin there and will weaken the key. Further, plugs cannot be fitted too closely to the cylinder lest they lock because of thermal expansion differences, dirt, and grime. The same holds for the pins and drivers; some allowance must be made for wear change and normal manufacturing tolerances. However, there are many cities in this country that have far fewer than forty thousand inhabitants, so the chance of anyone having a key duplicate to your own is very slim.

Pin tumbler mechanisms are used in padlocks, cabinet locks, case locks, and locksets, to mention a few variations. Figures 29 and 30 illustrate some of the pin tumbler applications possible, other than locksets, which are discussed in the following chapter. Locksets are economy locks made by combining a pin tumbler lock with a pair of doorknobs and a spring latch.

29. A four-pin tumbler filing cabinet lock. It has an overall length of two inches and is less than one inch in height. Its dead bolt is seen in the extended position. *(Courtesy Chicago Lock Co.)*

30. A drawer lock of polished brass. It has four pin tumblers and can be had with a dead bolt or a spring latch bolt (not shown). *(Courtesy Chicago Lock Co.)*

TYPES OF CYLINDERS

The pin tumbler lock constructed for use on a door is a cylindrical unit, complete in itself. It can be readily removed for repair or replacement without disturbing the rest of the lock mechanism. There are two basic types. One is called a rim lock, or rim cylinder. It is fastened to the door, or what have you, by two bolts connecting to its rear (see Figure 31). Sometimes the cylinder sits flush on the door. Sometimes, when the door is too thin for that cylinder, a washer or bezel may be placed on the exterior of the cylinder. Some bezels made for this purpose are adjustable.

The other type of pin tumbler cylinder is called a mortised cylinder, or a lock cylinder. It is identical in design and construction to the rim cylinder, with one exception. The mortised cylinder is threaded on its exterior. Fortunately for us, lock manufacturers, unlike plumbing-fixture manufacturers, have agreed on one common thread for all mortised cylinders, so that in most cases they are simple to replace.

To prevent the cylinder from simply being unscrewed from outside the door by a burglar, there are slots cut across the thread on

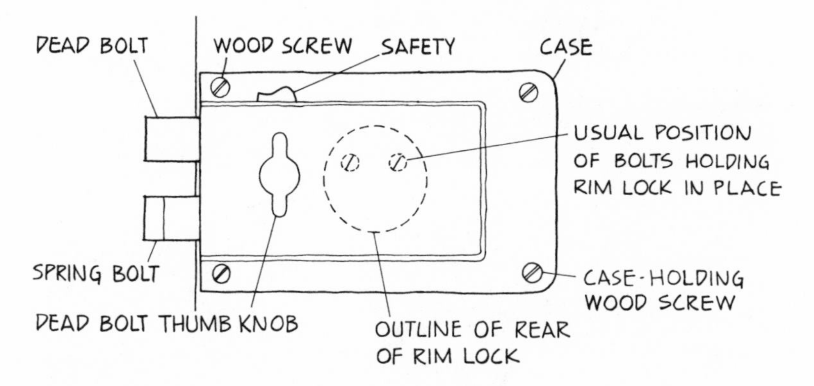

31. Simplified rear view (inside house) of a rim lock. To get at the bolts holding the rim lock in position, one must remove the lock case, which is done by loosening the four screws visible from this side.

the outside of the cylinder. Set screws mounted in the lock engage these slots and prevent the cylinder from being turned (see Figure 32). As further protection, armor plates are often used to keep the set-screw heads out of easy reach. Thus, to get at the set screws, one has to open the door, remove the plate, and then back the screws off.

INSTALLING A RIM LOCK

Figure 33 illustrates a rim lock in position and the measurements that may be used to locate its mounting hole. Only one is required with most locks of this type.

The first step is to measure the distance from the face of the lock to the side of the cylinder (A). The second is to measure the diameter of the cylinder (B). The next is to find the center of the cylinder (C). This may appear to be a complicated way of doing the job, but it eliminates the problem of transferring the edge of the lock up to the front plane of the cylinder, a procedure that requires careful use of a square or a plane working surface. Once the aforementioned distances have been ascertained, half of B is added to A to secure D. This is the distance that must be measured off from the edge of the door to the center of the cylinder hole that is to be

32. Typical cylinder constructed for use with a mortised lock. The cylinder is screwed into the lock. The set screws are then tightened, and they prevent the lock from being unscrewed.

cut. Generally, rim locks are mounted about chin high so that they are easy to see and operate.

There is no trick to cutting the hole for the cylinder other than taking care to make certain it is not so large that the entire lock, rim, and possibly bezel will pass through. A bit is chosen to cut a hole a fraction of an inch wider than dimension B. The best tool is a large wood bit. Lacking this, one can use an adjustable bit. When the point of either type of bit comes through the door, stop cutting and start from the other side. This will insure a clean-edged hole.

The cylinder is then passed through the door. Its back should fall a little short of the inner surface of the door. If it doesn't, use a washer or bezel or get a lock that is shorter in length. If it is longer than the thickness of the door, it cannot be tightly bolted unless a plate is cut and placed on the inside of the door, which is more of a nuisance than using a bezel.

The cylinder is held in place with one hand while the balance of

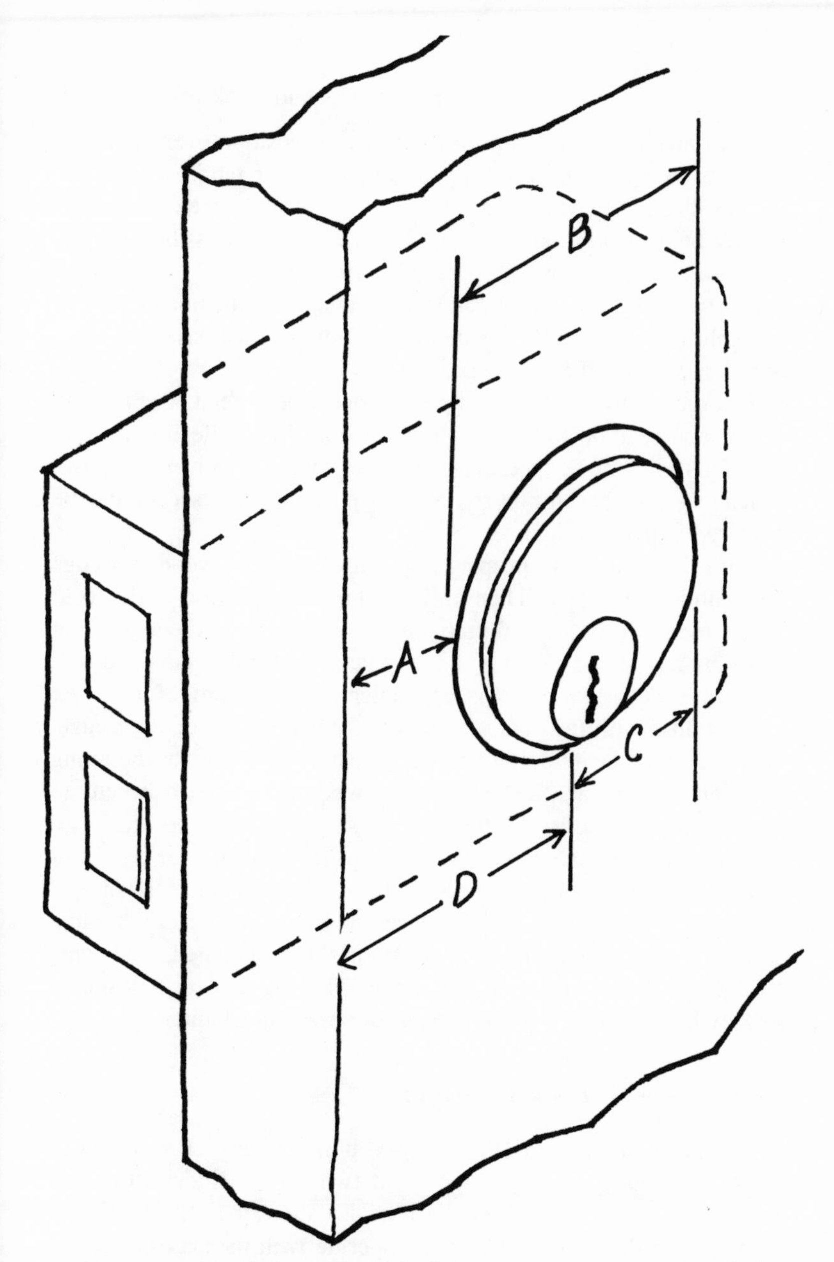

33. How to locate the center of the hole necessary to mount a rim lock. Measure A on the lock itself. Measure the diameter of the cylinder (B). Take half of this distance (C) and add it to A to get D. Now measure D from the edge of the door, set your drill, and cut the hole a little larger than B.

the lock mechanism is temporarily placed in position over the cylinder's inside end. The cylinder will have a tail, or tang. In some instances this tang will be a short length of flat steel, somewhat loosely held in position. That is to say, it will be possible to raise and lower the tail a bit (Figure 34). In addition, this metal strip may be so long that the inner lock mechanism will not fit over it and still lie flush with the inside of the door. In such cases the tang, or tail, is shortened as required.

The reason for the flexibility is the occasional need to offset the cylinder in relation to the lock mechanism. With a flexible tail the cylinder need not be in exact alignment with the lock mechanism. Whenever possible, however, the two parts of the lock should be mounted in perfect alignment.

With the tang cut to proper length, use a chip of wood to wedge the cylinder into place. Then position the inside portion of the lock, which may consist of a thumb knob that operates the dead bolt from the inside and possibly a safety latch that holds the dead bolt in a withdrawn position for convenience. The front of the lock mounts flush with the door, its body at 90 degrees to the door edge. With the lock in its correct position, carefully mark its mounting holes and remove it. Next remove the wood chip from between the cylinder and the door. Put the cylinder's back plate in position, and run the cylinder bolts home. Make them snug but not so tight that the metal plate is deformed. Next the lock mechanism is slipped over the extended tang, and the three or four wood screws holding the lock are screwed into their pilot holes and tightened. Following this, the strike plate may be mounted. The strike plate is installed exactly like all other strike plates, as discussed in Chapter 3.

THE MORTISED LOCK CYLINDER

The cylinder used with the mortised pin tumbler lock is similar to the rim cylinder in every respect but two. The cylinders for mortised locks are threaded on the outside, and generally they have cams rather than tangs. The cams operate with mechanisms set to one side of the cylinder rather than behind it, as in the case of the rim cylinder.

Figure 35 shows the innards of a mortised pin tumbler lock mechanism—without the cylinder. The lock shown has a dead bolt,

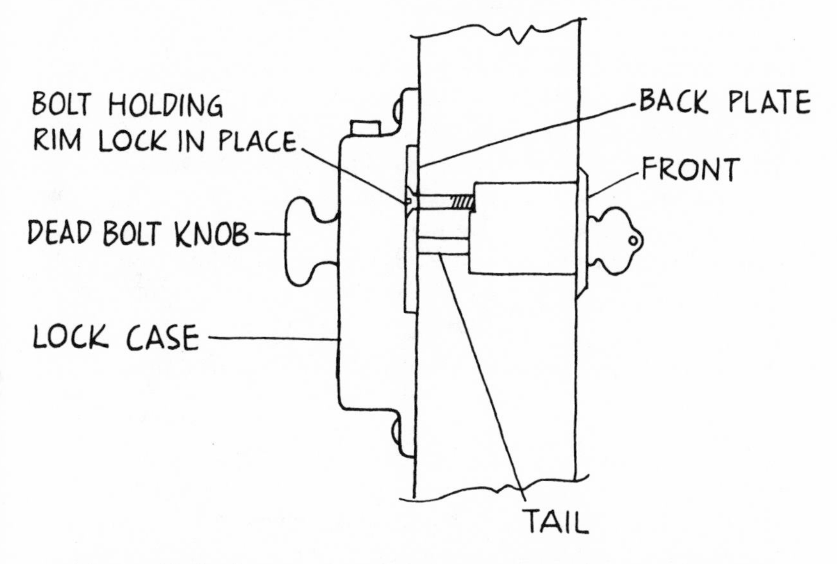

34. Simplified side view of rim lock showing how tail or tang may tilt a bit to accommodate the dead bolt mechanism on surface-mounted case inside the door

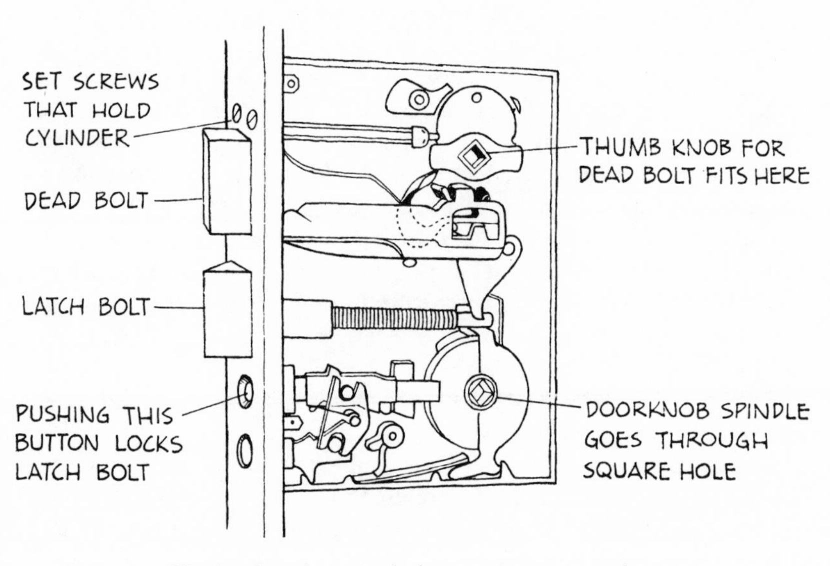

35. Interior of mortise lock case, cover removed

a spring latch bolt, and safety buttons. Since they are the simplest in operation, they shall be explained first. Note the square hole for the shaft holding the door knobs. Note the levers attached to the shaft mount. One lever activates the latch bolt. Turning the knob retracts the latch bolt; releasing the knob lets the spring behind the latch bolt drive the latch out and into the hole in the strike plate. Note that the two push buttons actuate one lever. One button, upon being depressed, sends the lever inward; the other sends it outward. Although it cannot be seen, there is a notch in the metal semidisk surrounding the spindle hole. When the lever engages this notch, the doorknob spindle is effectively locked and the doorknobs—either one of them—cannot be used to retract the spring bolt.

The dead bolt is connected via a series of levers to a cam mechanism that engages—in this design—a square tang fastened to the end of the plug on the cylinder. To facilitate opening the door's dead bolt from the inside, as well as the latch bolt, the upper square hole—the one connecting to the end of the plug—is also connected to a thumb knob that appears on the inside of the door. Turning this lever does two things. It can extend or withdraw the dead bolt, and it can withdraw the latch bolt. The latch bolt shoots forward in response to its spring.

The cover plate for this lock is shown in Figure 36. The cover plate carries the threads that hold the cylinder in place. It also holds, in this design, one long set screw used to lock the cylinder in place. The other set screw mounts on the case of the lock. The heads of both set screws are visible when the door is opened and one faces the front of the lock. Since backing these two screws off a quarter of an inch makes it possible simply to unscrew the entire cylinder from the front of the door, better installations have a safety plate that mounts over the face of the lock. Thus it is necessary to unscrew the safety plate before the locking screws are accessible.

INSTALLING A MORTISED PIN TUMBLER LOCK

Basically the same procedures employed to install a mortised warded lock (see Chapter 3) are used to install a mortised pin tumbler lock. The holes are laid out and cut the same way. The escutcheon is also installed in somewhat the same manner. The only

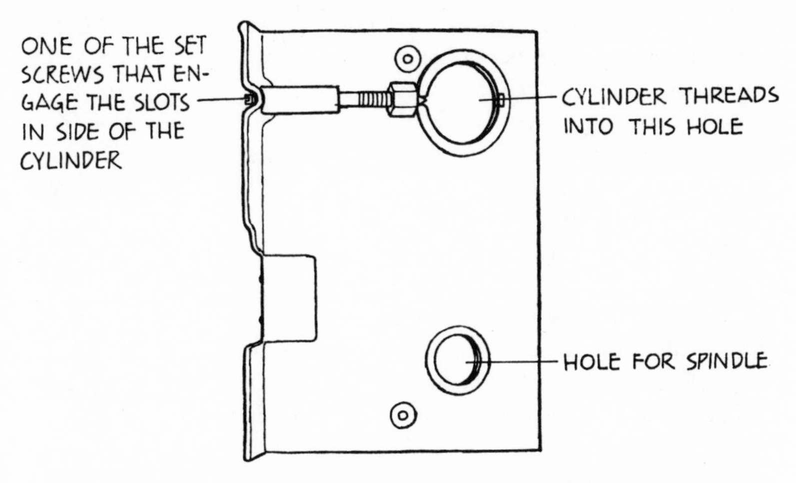

36. Cover to case in Figure 35. When mortised lock is screwed into threaded hole, its tang enters square hole, the other side of which takes the thumb knob controlling the dead bolt from inside the building.

difference is that the cylinder must be of a length to suit the thickness of the door and properly engage the dead bolt mechanism. In the illustrated lock (Figure 35) the cylinder's plug is fitted with a square-ended tang. This tang must be lined up properly. If the lock is mortised in place first, then it is a simple matter to locate the front and rear escutcheons by installing the cylinder and thumb bolt respectively first, with the escutcheons hanging on them.

To remove the cylinder the two set screws are loosened and a key is used to turn the cylinder out. If it has corroded itself tight, you can use a pipe wrench on it. This sounds like an exercise in brute strength, and it is; but there is no simple way of getting a rust-cutting solvent to the cylinder threads, and a length of steel forced into the keyway will most likely damage it irreparably. The marks left by the pipe wrench on the edge of the cylinder can be filed smooth and polished.

WAYS OF MAKING KEYS TO FIT A PIN TUMBLER LOCK

There are five ways to make a key to operate a pin tumbler lock. (1) Duplicate an existing key. (2) Make the key from the lock's code number. (These numbers are restricted; many recognized, established professional locksmiths get them only by cooperation with other locksmiths and cannot always secure them from the factory.) (3) Take the plug out and cut or rearrange the pins to fit the key. (4) Insert a blank key and use pin height or depth as a guide to cutting the key to fit the plug. (5) Leave the plug in its cylinder and cut the correct blank to fit by trial and error.

The first method requires that one have the correct key to duplicate. The other four methods will produce keys without the correct key in hand. The last method can be used to make a key for a lock that is still in the door. None of the methods give the expert much pause. Even making a key fit by trial and error will take the expert, with a vise to hold his key, no more than five minutes.

DUPLICATING A KEY

Duplicating keys by hand is a time-consuming but relatively simple task. First one must secure the correct key blank. These may be purchased at a locksmith shop, and at this point one is better advised to have the smith make the key, as this can be quickly and cheaply done by machine.

There will be times, however, when one has the correct blank or a key that can be further cut to match an existing key. In such cases the procedure is simply to "mate" the keys, side by side, in a vise and file away. This is illustrated in Figure 37. Care must of course be exercised to keep from cutting too deeply, and it may be advisable to color the old key with candle smoke or the like so that it is easy to keep from cutting the old key when cutting the new one alongside it. It is also advisable to make the cuts somewhat shallower than those of the old key at first. Try the new key from time to time, after filing the burrs away. It is possible that the old key has worn down a bit and that better, smoother lock response can be secured with higher bitting.

REMOVING THE PLUG

Methods 3 and 4 require that the lock's plug be removed from its cylinder. This is just routine when an operating key is on hand or

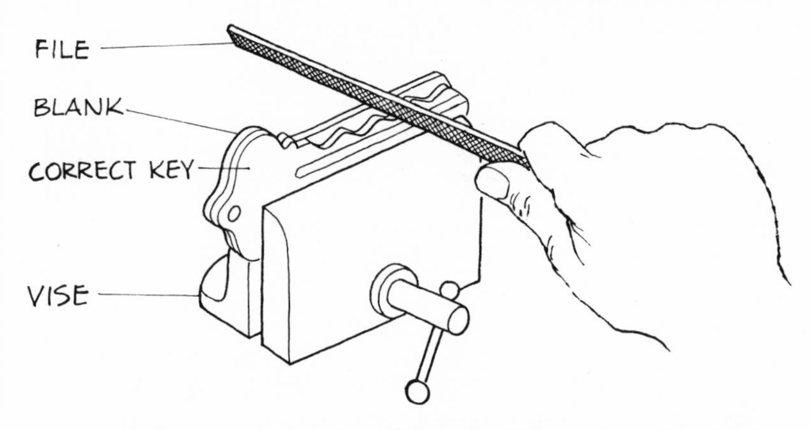

37. Making a key without the aid of a key-making machine. The correct blank is aligned next to the correct key in a small vise. Then a fine file is used to cut the blank. It requires care but is not difficult.

the lock has a cap over its drivers and springs. If one doesn't have the correct key and the cylinder doesn't have a cap, the task of removing the plug from its shell can be difficult to impossible. In any case, the first step is the removal of the cylinder from its associated lock mechanism.

Next, the two screws holding the cam or tang in place are removed. Then the cam is removed. A couple of clean saucers or clean plastic cups will be found very helpful for temporary storage of small parts. To hold the pins and drivers, one devises a "layout board." This can be easily made from a length of styrofoam or soft wood. It is simply a length of wood with a number of fairly deep parallel slits cut into it. You will need one slot for each pin hole in the lock you are working on. The slots should be numbered. If you are right-handed, calling the first to the left number 1 and so on is probably simplest to remember. For safety, it is good practice to mark "lock front" on the extreme right-hand side of the board (see Figure 38). The purpose of the system is merely to keep the parts neatly in correct order so that the pins, drivers, and their associated springs go back in the proper sequence, each group of three parts returning to the well whence they came.

Assume that the cylinder has a cap over its pins. This is the sim-

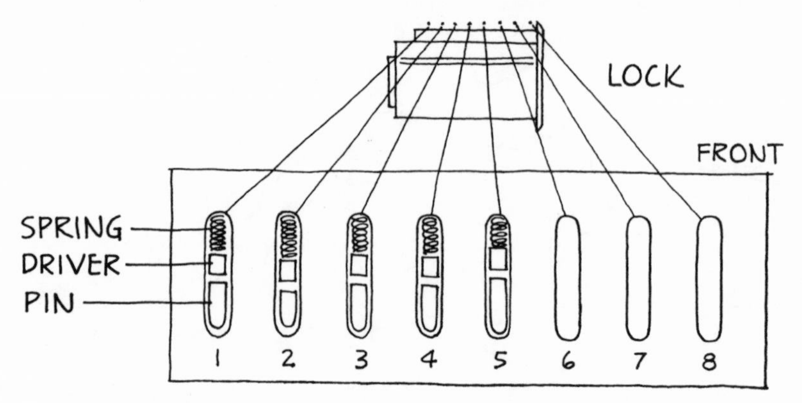

38. A layout board made by cutting a number of grooves in a length of plywood or Styrofoam is of considerable help in keeping pins and drivers in their proper relation.

plest situation, so we will discuss it first. Using a small screwdriver or a small pair of short-nosed pliers, push back the cap over the pins. As each hole is uncovered—don't uncover them all at one time if you can possible help it—the parts are placed in their corresponding slot in the layout board. When this has been done, the plug will come out.

The plug is now placed in a thimble or a small vise, pins carefully held right-side-up so they do not fall out. If the cylinder does not have a cap over its wells (or you may wish to use a key, which is easier, if you have experience) and you have the correct key, it is inserted and the plug is given a partial turn. Bear in mind the fact that the cam is off; the plug can easily fall forward and out. Also, *the drivers and springs are still in place, and they too will fall out.* Do not let this happen. Do not move the plug forward and out until you have backed it up with a following tool. A following tool consists of a rod of metal or wood that is somewhat smaller in diameter than the plug it follows and has a notch on its front end conforming to the rear end of the plug (see Figure 39).

There are ways to remove a plug from a cylinder without the aid of a key or access to the cylinder's springs and drivers through a covering cap. Three of these methods will be discussed here; the fourth, picking the lock so that the plug can be rotated, is discussed in Chapter 10.

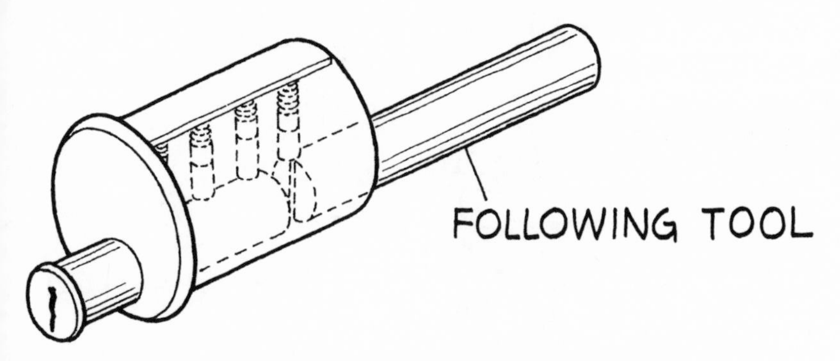

39. The following tool is used to keep the springs and driver pins from falling into the cylinder shell when the plug is removed. In practice the tool would be tight against the end of the plug.

With the first method, begin by wrapping some kind of tape over the cylinder's threads if it is a threaded cylinder. Next, insert a key blank or a tension wrench into the keyway. A tension wrench is described in the Glossary. Hold the cylinder in your left hand (if you are right-handed). Use your thumb to place tension on the tension wrench or on the key blank if that is what you are using. The tension's direction is such that it will turn the plug in the shell, right or left; it doesn't matter which way, though some plugs will go more easily in one direction than the other. Now take a wood or plastic mallet or a length of hardwood in your right hand and rap the front of the lock as hard as you can (Figure 40). Given a clean moral record, a healthy attitude, and a little luck, you may "crack" the cylinder with the first whack. If you have been wicked, you may blue your thumb well before the lock will crack open. Don't punish yourself too much; some locks will not crack open this way even for experts. Try another approach which, though best with old and worn locks, may work with a new lock.

Hold the lock in the left hand, two fingers to either side of the plug and the thumb pressing inward on the rear of the plug, as shown in Figure 41. Hold the front two fingers slightly across the face of the plug so that the plug will not shoot out should you crack it. Now rap the side of the cylinder with a mallet. Experts hold the plug in their right hand and rap the edge of the cylinder against a

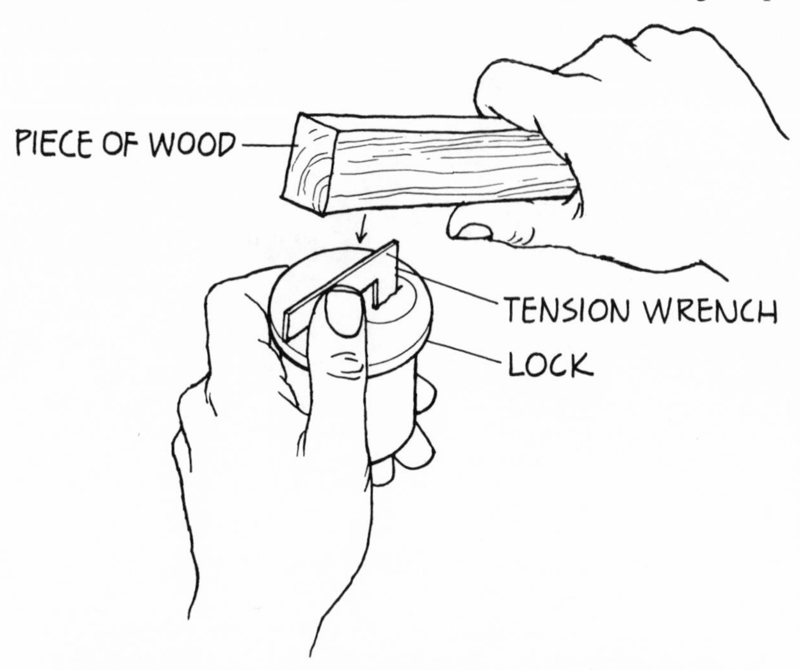

40. One method of cracking a cylinder. The thumb is used to apply a rotational pressure to the plug via the tension wrench while the face of the cylinder is sharply rapped with a length of wood or mallet.

hardwood table top. Again, there is no point in repeating this effort endlessly. Some locks, especially well-worn locks with ball bearings below the pin, will not crack open this way. The balls hang up in the worn grooves on the sides of the pin holes.

The third method of keyless plug removal involves the use of a long length of very narrow, very thin steel spring. Generally a length of watch spring is used. If it coils too severely, it can be annealed somewhat in a gas flame.

The cylinder is removed from the lock, and the cam is removed from the cylinder. A blank key is inserted in the keyway; then the steel strip is inserted along the shear line through the rear of the cylinder. With one hand the blank key is gently moved in and out, up and down, while the steel is gently pressed inward. The trick is to get the steel between the pin and its driver, working forward from

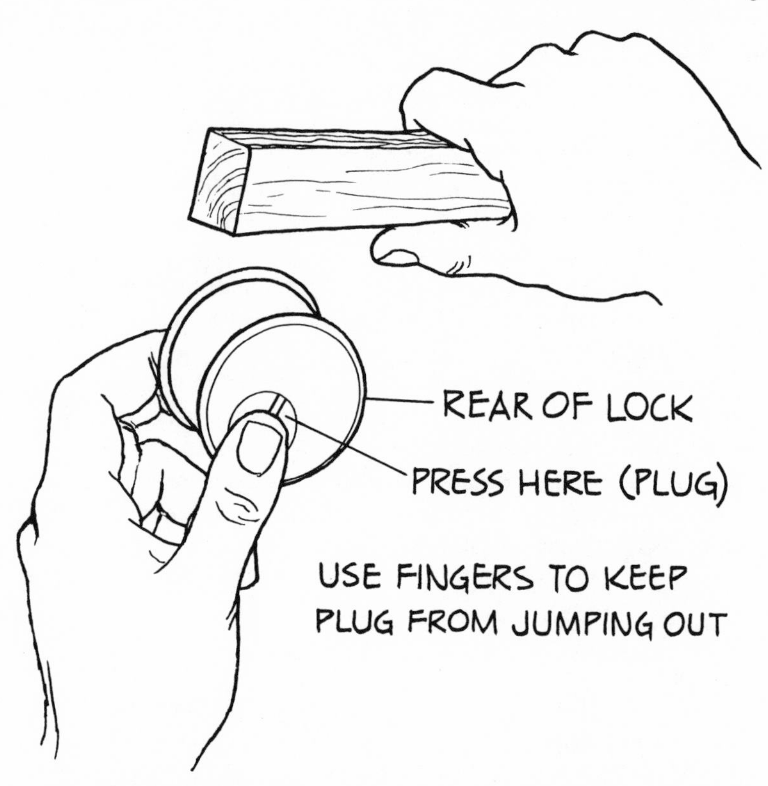

41. Another method with which to crack a pin tumbler lock. After the cam has been removed, the thumb is used to apply pressure to push the plug out. The length of wood is used to strike the side of the lock—not the fingers—to shock the pins into the correct position. Hold fingers over front of lock to keep plug from spilling its pins on the floor.

the rear end of the plug to its front. This can be done if the steel is slim enough and there is sufficient space between the plug and its shell. Obviously, there is a much better chance of accomplishing this with an old and worn lock than with a new one. Sometimes, too, sloppy workmanship on the part of others helps. Some smiths file too vigorously when cutting tumbler pins down. As a result the shear-line clearance is increased.

Although all three methods work, they don't always work, even in the hands of experts. The one remaining means of removing a plug without a key—picking a lock—doesn't always work either. Still, with hard work and practice, almost all locks can be opened one way or another. Experts can pick a lock in a minute or two and crack open a worn lock with one slap against a bench top. Not everything has given way to the machine and computer.

In any case, the plug is now out of its cylinder and neatly ensconced in its thimble or lightly held in a vise. The shell is off by itself. Its drivers and springs are either held in place by a following tool, or, if the cylinder had a cap, the drivers and their springs are in their proper positions in the layout board. The next step is to insert a key into the plug.

If we are going to cut or rearrange the pins to fit a specific key, that key is inserted. If we are going to cut a key, whether it is a blank or an old key, that key is inserted. No matter what the choice is, the key inserted must have the correct cross-section. If it enters easily, it has the correct cross-section. Stated in locksmithing terms, it is the correct blank—or has been cut from the correct blank.

CHANGING A PLUG TO FIT A KEY

Let us assume that we want to make the plug (actually its pins) fit the key. Figure 42 shows a hypothetical key in our plug. Since the key has entered the keyway, we know its cross-section is correct, but to check its length we look down the end of the keyway. If its end is beyond the rearmost pin, it is long enough for the cylinder. If it is too long, it can be shortened.

Remember, in the present example the key is the master: The pins must be adjusted to accommodate the key. High pins must be lowered, short pins must be increased in length.

The first step is to take a pair of tweezers and shift the pins around till they are all flush with the top of the plug or above the shear line. None can be used that fall short of the shear line. If no pin shifting can accomplish this, then the remaining short pin or pins must be replaced with longer pins. These may be taken from another lock. Any pins of the correct length and diameter can be used. Obviously they should be of mild steel and not so thin that they can fall crosswise and jam or so thick that they stick in the hole.

The high pins are lowered by filing them with a flat file while

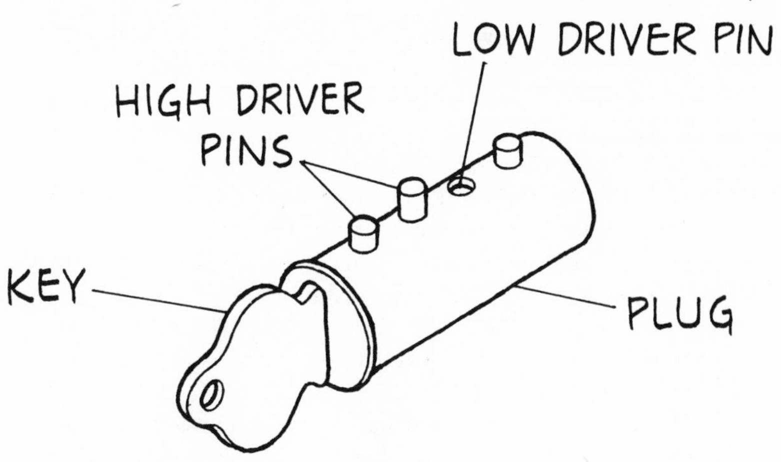

42. The key to be fitted is in place in the removed plug. Note that in this example there are three pins projecting above the shear line and one below the shear line.

they remain in place in the plug. Some tumbler pins have rounded tops. If this is the case, for best results each pin that is to be shortened should be removed and its top rounded. The high spot on the rounded top must be flush with the shear line. When finished, the plug and its pins should be carefully brushed to remove metal particles. Some shops wash them down with a little gasoline to remove all grease traces and dirt particles.

CUTTING THE KEY TO FIT THE PLUG

When the key is to be cut to fit the plug, the plug is the master. Its pins cannot be moved from hole to hole, nor their lengths altered. If this is done, the newly cut key will fit the lock, but all other keys will be "locked out."

The key to be cut is inserted. Again it is checked for fit and length. If it is a blank, a new key that has not been cut, all the tumbler pins will be raised flush with or above the shear line. If they are not at least flush, something is wrong with the blank or someone has tampered with the lock.

If the inserted key has been previously cut, all the tumbler pins must rise to flush or above. If any cut in the key is so low that a pin doesn't come flush, that key cannot be used with that lock.

It is now necessary to cut the key so that all the pin tops lie flush with the shear line. This may appear difficult. It is not; it is merely time-consuming. Take a pair of machinist's dividers and lay off the spaces between the pins on the bit edge of the key. In other words, transfer the imaginary line running down the center of each pin to the bit edge of the key. This will mark the areas that may need to be filed (see Figure 43). Another way to do this is to blacken the key with soot from a lighted match or candle. The key is inserted into the plug and moved a little. Bright spots indicate the points that contact the pins.

By looking at the distance each pin projects above the top of the plug, one can judge how much filing is necessary below each pin. Filing is done slowly to make certain that one does not overfile. After all the cuts have been made, the sides of the key are filed with a flat file to remove burrs and whatever sharp breaks between pin areas exist. This is done so that the key can enter and leave the keyway without catching on a pin.

DRIVER-PIN LENGTHS

When a key is cut to fit a particular plug, as just described, there is no change in the parts of the cylinder; all pin lengths are untouched. When a plug's pins have been altered or moved to accommodate a particular key, length changes have most likely been made.

If the total height of a tumbler and its driver plus the top of the key is greater than the total space provided, it may be impossible to insert the key all the way, it may be difficult or impossible to remove the key once it has been inserted, or a spring or springs may be crushed, affecting the operation of the lock adversely.

With experience, it is a simple matter to look at the changes made to a particular plug and its pins and know if sufficient space exists to accommodate the change. Without experience it is best to make certain. To do so, a beginner is advised to remove the springs and drivers and place them in their correct layout-board positions. It is also advisable to disassemble the shell and its parts for the pur-

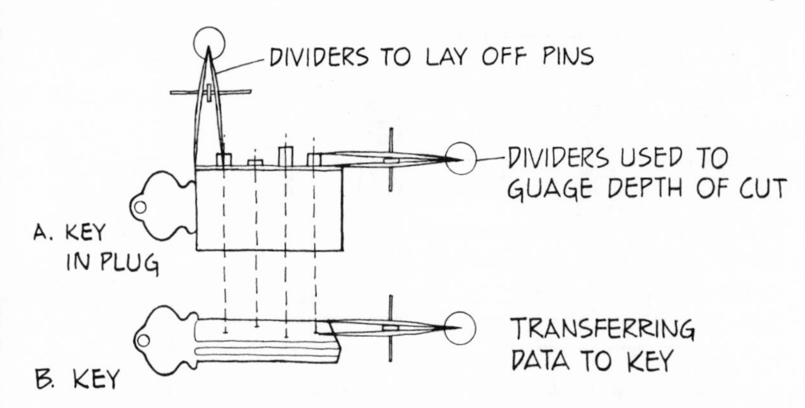

43. How a pair of dividers can be used to locate the areas that need to be cut when fitting a key to a plug. As cuts are made, key is repeatedly tried in plug.

pose of inspection and cleaning when the plug is out unless the lock is new and obviously in good condition.

The plug is put aside for the moment. The shell, with the following tool in place, may be placed in a vise for convenience. Next, the following tool is slowly withdrawn. As each driver and its associate spring falls down, it is carefully removed and placed in its proper slot in the layout board.

The pins are now removed from the plug and placed in their respective slots in the layout board. Using a straightedge (a small ruler will do), push the pins and their drivers together to form a horizontal line. Inspection will quickly reveal which pair of pins and drivers is longest (Figure 44). At the same time the parts should be inspected for grease and grime and the springs inspected for crushing and deformation. These must be replaced with springs of equal size and tension. Cleaning may be done with white gasoline or kerosene. Do not use Carbona (carbon tetrachloride), because it is a poison. Do not use turpentine or similar solvents, which may leave a residue.

Return to the cylinder shell. Give it half a turn and tighten the vise. Its wells are now at the bottom, the reverse of its normal operating position. Return to the layout board. Pick up the longest trio of pins, drivers, and springs and insert the spring in the well nearest

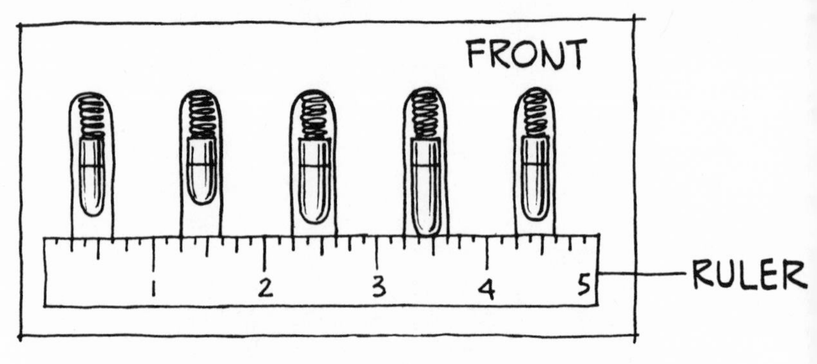

44. Using a small ruler to find the trio of springs, drivers, and pins that are longest on a layout board. They are used as a guide.

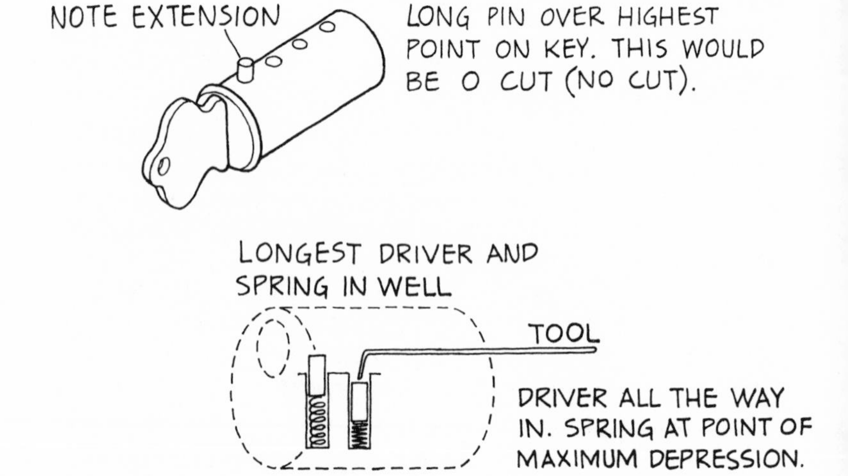

45. Testing the maximum overall height of a pin, driver, and spring to see whether or not there is enough space in the pin well to accommodate them

the front of the shell (see Figure 45). Now place the associate driver atop the inserted spring.

Note how much of the driver projects into the plug hole. Depress the driver with a small screwdriver. Can the entire driver be pushed easily into its well? Now take a bent nail or wire and depress the driver a little more. How much play is there before the spring "bottoms" and can't be depressed any more? Is there enough space to accommodate the long pin? To see exactly how much space is needed, place the long tumbler pin in any hole in the plug. Insert the key and see how much of the pin extends beyond the plug. The long driver and its spring must accommodate this much tumbler pin.

If there isn't sufficient clearance, the long driver can be shortened by filing, or it can be replaced with a shorter pin. The tumbler pins, the pins that go in the plug, cannot be changed.

If all this seems like a lot of fuss to determine that ⅛ inch or so of metal must be removed, remember that it is quite possible to lock a cylinder of this type forever. If the overly long pin jams, there may be no way of removing it.

SETTING UP A CYLINDER

Now that the drivers and their springs have been removed, inspected, and cleaned, they need to be replaced. This is called setting up. First, however, if the pins are dirty, the entire shell should be meticulously cleaned. Use pipe cleaners and a hydrocarbon cleaner. *Do not lubricate the pin tumbler mechanism with oil or grease of any kind.* The only suitable lubricant is graphite. If powdered graphite in a graphite gun is not available, rub the parts with a soft lead pencil. A trace is sufficient. Too much will clog the works.

Some lock makers drill all their locks for six pins but divide their line into two groups. The better locks have the sixth pin and driver. The others have merely the empty well.

If you find a lock with an empty well, do not install a driver and spring. If you do, none of the other keys cut for this particular lock will work.

At this point we have a clean plug with the correct pins inserted in a thimble on our work bench, and a clean shell with clean and tested drivers and springs in the layout board. The next step is to return the springs and drivers to the shell and insert the plug. In other words, we are now ready to reassemble the cylinder.

The shell is still in the vise, pin holes down. Remove the test driver and spring, if you have not already done so. Take the following tool and insert it into the rear of the shell, notched end forward. Insert the tool halfway. Take a pair of tweezers and insert spring number 3 or 4 into its proper well. Follow this with the proper driver. Push the tool in a bit to hold the driver down. Insert the next spring and driver and push the following tool forward again. Do this until all the wells open between the front of the following tool and the front of the cylinder have been filled. Now carefully pull the following tool through the shell until the empty well near the center of the cylinder is exposed. Fill it and move the tool rearward a bit. Fill the next hole and so on until all are filled. Working from the center out is not the only way of setting up a cylinder, but it eliminates the long reach down the length of the plug hole (see Figure 46).

The following tool is now in place. All wells have been correctly filled. The shell is next loosened in its vise and given a half-turn so that the drivers are on top. Now the plug, pins in their proper place, is turned a fraction to the right or left and slowly backed into the slot in the following tool and backed into place. Maintain a slight pressure between the plug and the notched end of the tool so that

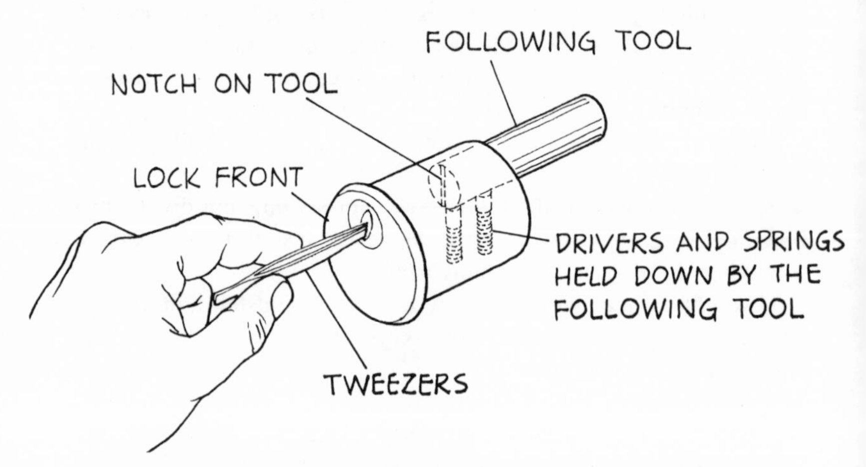

46. The following tool is used to hold the springs and drivers in place as the cylinder is set up. After the wells have been filed, the shell is turned to place the drivers on top before the plug is inserted.

no space develops and so that no pins can jump into this space. After the plug has been completely inserted, its cam is installed.

There are two important details to bear in mind at this point. Failure to remember them can lead to unhappiness. Make certain that the notched end of the following tool sticks out the front end of the lock. Make certain that the plug is slightly turned—that its pins are not vertical and in a direct line with the driver holes.

If the above instructions are not followed, it is possible for a pin to jump out and into the space between the tool and the end of the plug. If the plug holes are in line with the case holes and the first driver or the second driver jumps down into the pin holes, the lock will be jammed, and it may be impossible to get the plug out again without picking the lock.

Some lock manufacturers install steel balls beneath their tumbler pins. They usually can be seen from the front of the lock. If they are used, they must be left in place. A not-too-good alternative is to replace all the tumbler pins with longer pins of the correct length when the balls are removed for one reason or another.

MAKING A KEY BY TRIAL AND ERROR

Keys can be fitted to pin tumbler locks without removing the lock from the door and without a correct key to duplicate. The beginner will find the process tedious and time-consuming, but an expert can make keys this way fairly quickly.

Again we start with the correct key blank. (It is difficult to impossible to fit a "cut" key this way to a lock that is in place.) The key is blackened by holding it in the flame of a match or candle. It is then carefully inserted and turned as far as it will go. Upon removal and examination bright spots on the key's bit will be seen. These indicate the areas that must be filed. The key is placed in a vise and a needle file is carefully applied. Very little metal is removed. The key is removed from the vise, blackened, and once again inserted into the lock and turned.

Again the key is removed and gently filed as indicated by the bright spots or scratches through the blackening. If no bright spot shows in one area, you have filed just enough, or perhaps a little too much. Obviously, since a few thousandths of an inch can be too much, the key will have to be blackened and inserted dozens of times before it is properly cut.

Experts use their knowledge of locks to judge the depth of the cuts necessary by "feeling" each pin with a pick, a pointed length of wire. And when an expert over-files, he increases the height of the metal at that point by hammering its side slightly to increase its height. The first time around, all these steps will be difficult, but with practice and an increasing "feel" for locks, they will become easier and easier.

These are the five general methods used to make a key for a pin tumbler lock. Notice that the use of wax to make an impression from which a key is to be cut has not been mentioned. Professionals do not use wax. For one thing, it can only be used to secure a duplicate of the key's cross-section. Since experts can select the correct blank by just glancing at the lock's keyway, they do not bother with wax.

TIPS ON SPRINGS, PINS, AND KEYS

In the foregoing discussion we have assumed reasonably good conditions: The springs were not damaged, the pins were of reasonable length and diameter, and the keys were properly proportioned and cut. This is not always true. In the course of repairing and setting up cylinders, one will encounter wrong parts, damaged parts, and poorly cut keys, and some parts will be missing. Let us consider the key first.

A correctly cut key is illustrated in Figure 47. Note that the cut nearest the bow is not too deep, and that a 0 cut is not followed by a maximum, or 9, cut.

Figure 48 illustrates two improperly cut keys. One key has an overly deep first cut. This weakens the key. The changes in bitting heights are too abrupt. This can cause difficulties. The key sticks in its keyway or is difficult to insert and remove. The second key suffers from regularity. All the bit cuts are alike. The plug set up for this key is most easily picked.

Pins, both drivers and tumblers, should be at least one and one-half times longer than they are thick, but preferably their lengths should be twice their diameters. If they are too squat they may cock

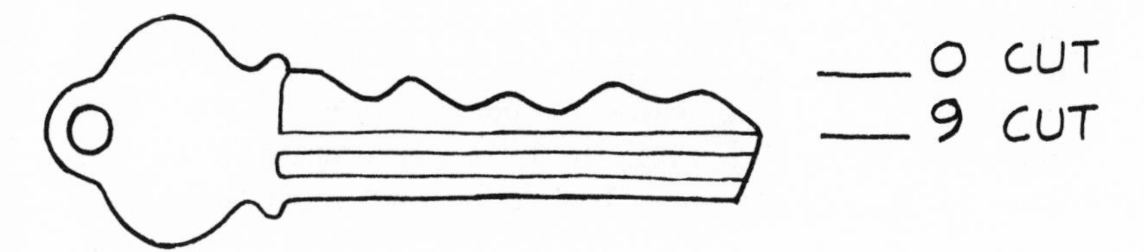

47. A correctly cut key. The cut nearest the bow is not too deep. There are no abrupt changes in the bitting.

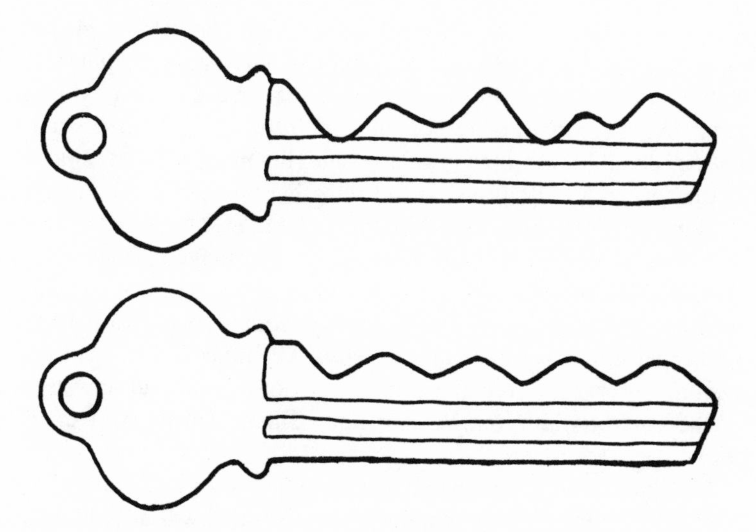

48. Two improperly cut keys. *Above:* The cut nearest the bow is too deep. The changes in bitting height are too abrupt. *Below:* Each cut depth is like its neighbors. The lock to this key is most easily picked.

in their wells and jam. When they stick, the lock becomes permanently open or closed and can only be picked with great difficulty. Pin diameter must be somewhat smaller than their wells, but again not so small as to permit them to get caught at an angle and lock into one position. They must be free to move up and down their holes.

Springs must be "alive" and show no evidence of deformation. Spring diameter must be small enough to enable them to move freely in their wells but not so small as to permit them to double up or jump sideways out of their holes.

Obviously, a good lock untouched by a "butcher" will have properly sized parts, but many locks carelessly set up or repaired will have poorly fitting parts. The repairman simply looked into his junk box for parts and used whatever could be fitted into the pin wells.

If you encounter a mixed bag of springs and pins, how do you recognize the correct or reasonably well-fitting parts? Close examination, possibly with the aid of a magnifying glass, will reveal kinks in the spring. Merely pressing and releasing the spring will measure its remaining resiliency. Now drop it into one of the pin holes. Does it fall readily to the bottom? Use tweezers to move it from side to side. Is there too much play? If the spring is broken or damaged it must be replaced, or the lock will only work in one position. The best replacement spring is that which is recommended by the lock's manufacturer. However, the springs are not critical and any spring of the correct width, length, and approximate tension will do the trick.

Pin diameter can be tested by inserting each pin in one of the holes and testing its play. If the pins are tight and will not fall to the bottom of the well when they are clean, replace them. Do not attempt to file them smaller or cut them down by rubbing them with emery cloth. The chances are that you will end up with an oval-shaped pin that may be loose in one position and tight in another. An overly narrow pin is just as bad. It may get caught on the wall, and it may slip inside the end of the spring.

Pin lengths can be easily checked. As stated, the shortest pin should not be less than one and one-half times its width, and preferably longer. Maximum pin length is governed by the depth of the pin well, the length of the compressed spring, and the total length of the driver and its associated tumbler pin.

Pins are best shortened by filing their ends. If there is too much to file, cut them with a fine-bladed hacksaw, but first place them between two boards in a vise. Do not clamp them directly in a vise or the pin's sides will be roughened.

TROUBLE-SHOOTING PIN TUMBLER LOCKS

The cylinder turns a little to the left or right when the key is turned. The cylinder is loose. If it's a rim lock, remove the lock mechanism from inside the door and tighten the two bolts holding the cylinder in place. If it is a mortised lock, tighten the two set screws running from the lock's face into the lock. Give the cylinder a partial turn to either side to make certain the set screws have entered their channel.

The correct key does not always work. If the position of the key in the keyway appears to affect its operation, see if there is any lateral play in the plug. If it can be moved out or in a fraction of an inch, the plug may be badly worn or the cam may be loose. If the cam is tightly in place and the play is in the plug, remove the cam and file the shoulders of the plug down a little bit. This will bring the cam closer to the rear of the cylinder. Replace the cam and try the key. If it no longer works or is difficult to operate, the key has been shifted in its relation to the pins. Either disassemble the cylinder to file the pins down, or otherwise correct them, or blacken the key and locate its high spots that way. If shifting the plug has permitted a driver to drop into the plug—with the key inserted—it will be necessary to replace that pin or pins with longer pins.

The correct key is difficult to turn. Try the thumb knob on the other side of the cylinder. If that too is stiff, and the door is open, remove the cylinder and try it again. If the thumb knob controlling the dead bolt is stiff, the trouble may be in the lock mechanism alone, in which case it is disassembled, cleaned, and greased. If the thumb knob is fine when the cylinder is removed (which also may be the case with the cylinder in place, depending on lock design), the trouble is in the cylinder. Do not take the cylinder apart before ascertaining that the trouble cannot be corrected by altering the key a little. Blacken it and test it. If that key must operate other locks too, it should not be altered. Disassemble the cylinder, examine it, and clean it.

The correct key will not open the lock or must be forced to make the plug turn. Disassemble the cylinder. It is possible that a spring has broken or pieces have fallen between the pins. It is possible that a short pin has worn so much that it is at an angle in its well or hole.

The plug turns but the dead bolt doesn't move. The physical connection between the plug and the dead bolt has been broken. If it's a rim lock and loose, tightening the holding bolts may do it. If not, disassemble the lock mechanism. If it is a mortised lock, the cam or tang has loosened from its shaft or some connecting lever has broken or moved out of place. Disassemble the lock mechanism.

The correct key will not go all the way into the keyway. If the key can be inserted and removed easily enough, there is a piece of broken key at the farther end of the keyway. If the key appears to become wedged partway in and needs considerable effort to remove, someone has attempted to force the lock with a screwdriver or steel bar and has mashed the keyway, making it narrower.

The piece of broken metal can be removed with a length of steel piano wire with a hook at its end or a broken-key extractor—a length of fret-saw blade, ground thin and attached to a handle for convenience. A mashed keyway can sometimes be opened with a slim file, but since this also opens the lock to possibly another key (different blank), it is not good practice. The lock should be replaced.

The wrong key opens the lock, in addition to the correct key. This is almost never due to a wear change in the cylinder but due to either master keying or poor repair of the cylinder. The second key may be a master key, cut to open this lock as well as others. Or the last person to repair the lock removed one or more of the pins, thereby simplifying the lock mechanism and reducing the total number of possible key changes. In other words, the lock has been made accessible to more keys. This can be corrected by replacing the missing pins and cutting the old key or keys to fit.

The key is difficult to turn when the door is closed but turns easily when the door is open. The door has moved or settled, or the door frame has warped or settled. The dead bolt rubs against the edge of the strike plate when the door is closed; therefore the key turns with difficulty only at that time. Raise door, file strike plate slightly as needed, and/or correct the change in the door frame.

The latch bolt sticks. This too may be due to a change between the lock and the strike plate. Open door partway and note where and how latch bolt meets strike plate.

The safety buttons do not operate. If they can't be moved, paint is the most likely cause. Remove and clean. If they can be moved but have no effect, an internal lever has moved out of place. A spring has broken.

The latch bolt will not extend by itself or at all. Look for a bent latch bolt—someone tried to force the door. Look for paint on the latch bolt, a broken spring, or a worn latch bolt. Sometimes adding a stop or bending the stop to keep the bolt from entering the lock so deeply that it catches on the case edge will cure this problem.

The thumb lever or knob will not retract the latch bolt fully. This may be due to wear in the series of levers that connect the thumb lever to the latch bolt, or possibly someone has hammered on the thumb piece, bending it or its associated levers. Replace or build up missing metal by brazing. Straighten bent parts with pliers.

The cylinder is position-sensitive: It can be operated only when held a certain way. This is a problem limited to padlocks and case locks. Cylinders on doors cannot be inverted. The trouble is a flattened or broken spring. The pins associated with this spring will assume their correct position only when they are above the plug. Sometimes tapping the cylinder makes the key work. Disassemble the cylinder; replace the broken or mashed spring. In the latter case correct the total pin length before replacing with a new spring.

The plug sticks a bit before responding to correct key. The key is worn. It was probably never perfectly cut in the first place. It might be improved by blackening it and recutting, but it is best to start with a new blank.

Chapter 7

Locksets and Other Locks

Lockset is a name given to locks that are combined with a pair of doorknobs and a spring latch that can be locked. Most modern homes have locksets on their exterior doors. The reason is simple economy. The lockset is easy to install and low in price. The better locksets utilize pin tumbler lock mechanisms; the lower-cost locksets rely on a disk tumbler lock for their security.

About a dozen companies are presently manufacturing locksets. While each lockset is basically the same as other locksets, there is considerable difference in construction details. Each company's products come apart and go together a bit differently. Some have antipick latch bolts; others do not. Some use brass knobs and parts; others use aluminum knobs and white metal castings. In general, the more expensive locksets offer greater security and greater wear. Again, very generally, the heavier locksets are usually better. They have heavier parts—more brass and less plastic and tin.

INSTALLATION OF LOCKSETS

The lock and knob portion of the lockset requires the drilling of only two holes: one large hole for the "works" and a smaller hole at right angles through the edge of the door for the latch bolt. Although machine-shop accuracy is not required for these two holes, goofing can ruin the door.

Use the bit size suggested. Do not use a larger bit or a smaller bit and plan on correcting the error later by one means or another. The large hole can be drilled with an expansion bit, but the fixed-size carpenter's wood bit, two or so inches in diameter, which most people have never even seen, will do a far better job and make a more accurate, cleaner cut. If you plan on installing more than one lockset, it pays to invest in a bit of this kind. You will have to go to a large hardware shop or a builder's-supply hardware shop.

Use the template provided with the lockset. Take your time and lay out the holes accurately. The center of the knob hole is usually thirty-six inches from the floor.

Figure 49 illustrates the installation of a Kwikset lockset. Other locks are installed somewhat similarly, but remember that each manufacturer provides his own template, and you can bet your boots they are all different.

LOCKSET REPAIRS AND MAINTENANCE

As far as the lock mechanism itself is concerned, the lockset differs little if at all from similar locks; however, getting to the lock is

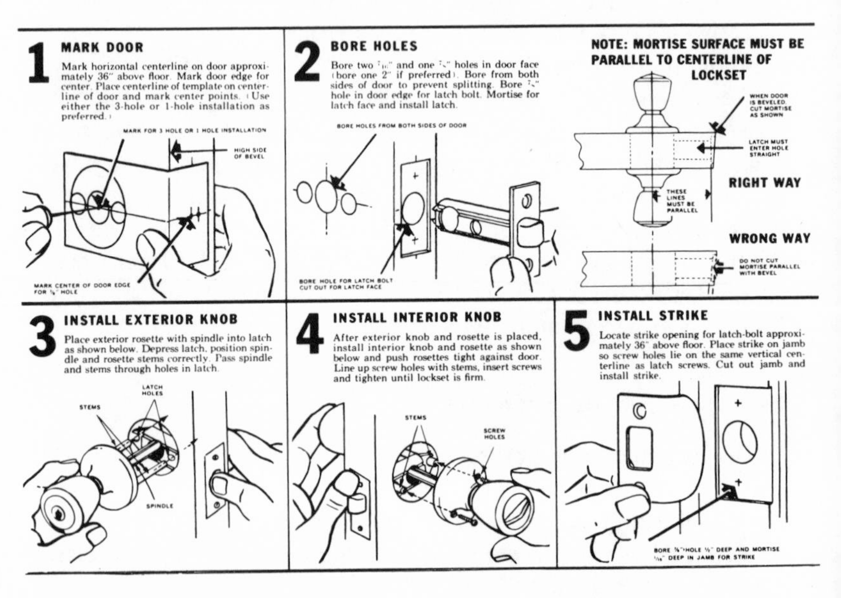

49. How to install a Kwikset lockset. *(Courtesy Kwikset Sales and Service Co.)*

different from the approach necessary with other types of locks and differs from one manufacturer to another. For example, the lock mechanism on the Vanguard line of Weslock locksets can be reached and removed without removing the entire mechanism from the door. Figure 50 illustrates how this is done to change hands (turn the lock 180 degrees so that the bitting on its key faces downward). The awl pictured in the illustration can be inserted into position after the lock's screws have been loosened enough to withdraw the collar. Figure 51 shows one of the Kwikset "400" line of locksets, and Figure 52 shows how the lock mechanism is removed.

Problems specific to locksets are relatively few. As a whole they wear more rapidly than other types of locks, for the knob sees all the action. When the knob and supporting bolt assembly wear, they can be replaced. The manufacturers stock the parts. It is this writer's contention, however, that it is best to replace the entire mechanism when this happens. Like the famous one-horse shay, when one part goes, they all go.

From time to time it is good practice to unscrew the two bolts and dust out the lock's interior. A few drops of fine oil on the spindle and locking mechanism help reduce wear. Do not drown the parts, or the oil may work into the pin tumbler.

OTHER TYPES OF LOCKS

No one has ever cataloged all the different types of locks invented, patented, or even manufactured. Some were too "far out" to be worthy of manufacture, others had basic defects that were not recognized until they were in actual use over a period of time, and still others failed simply because of a lack of showmanship: Their supporters did not have the flair of a Hobbs or a Bramah. Presumably a study of the patent records will show those locks that reached the patent office but went no further. If the demise of so many lock designs implied here appears exaggerated, think of the more than one hundred different automobile manufacturers that were actively in business in Greater Boston at the turn of the century. Where are they now?

One of the locks currently manufactured and in use but not discussed in this book is the Bramah. Variations on the original con-

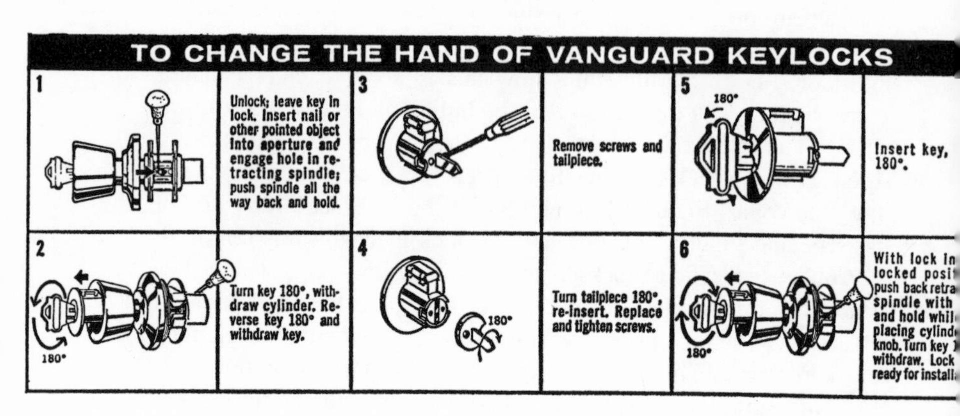

50. How to change the hand of a Vanguard line Weslock lockset *(Courtesy Weslock Co.)*

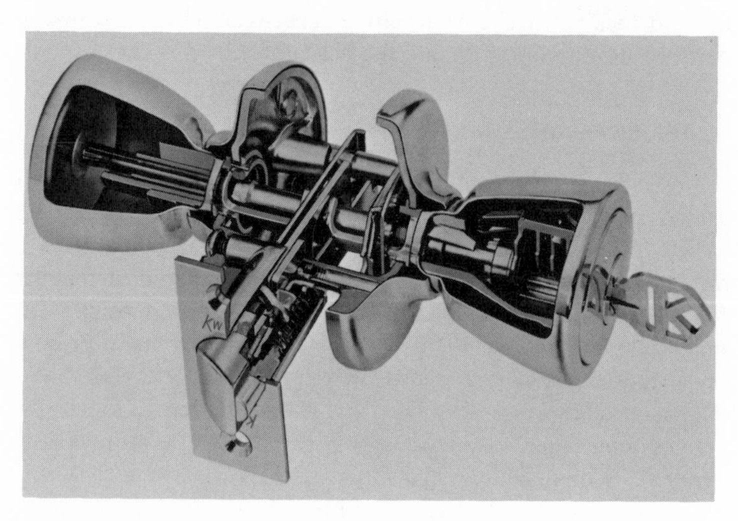

51. Cutaway view of typical key-in-knob lockset. This unit is one of the "400" line manufactured by Kwikset.

A. ALIGN SPINDLE

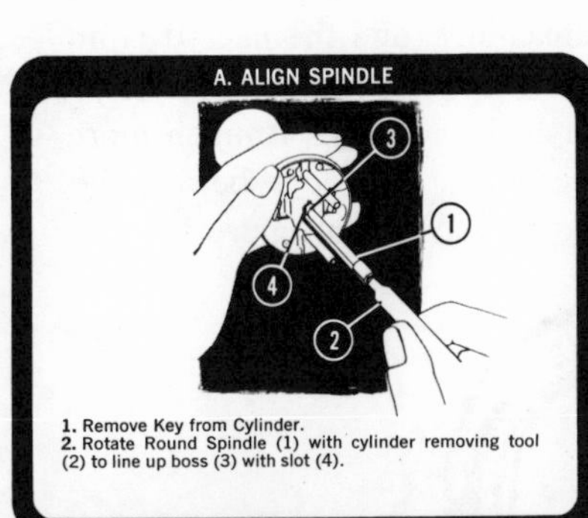

1. Remove Key from Cylinder.
2. Rotate Round Spindle (1) with cylinder removing tool (2) to line up boss (3) with slot (4).

B. REMOVE SPINDLE

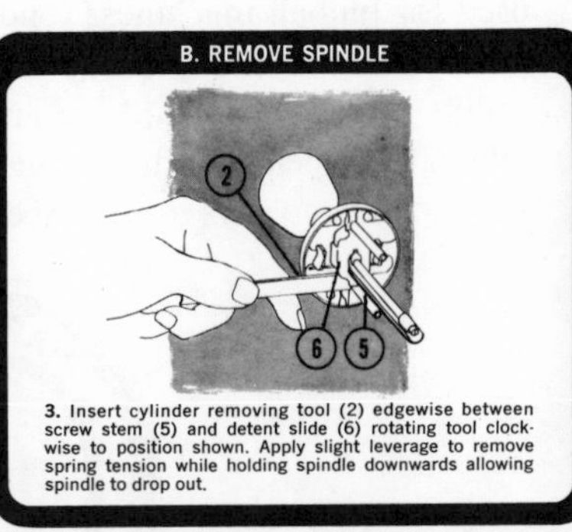

3. Insert cylinder removing tool (2) edgewise between screw stem (5) and detent slide (6) rotating tool clockwise to position shown. Apply slight leverage to remove spring tension while holding spindle downwards allowing spindle to drop out.

C. REMOVE CYLINDER

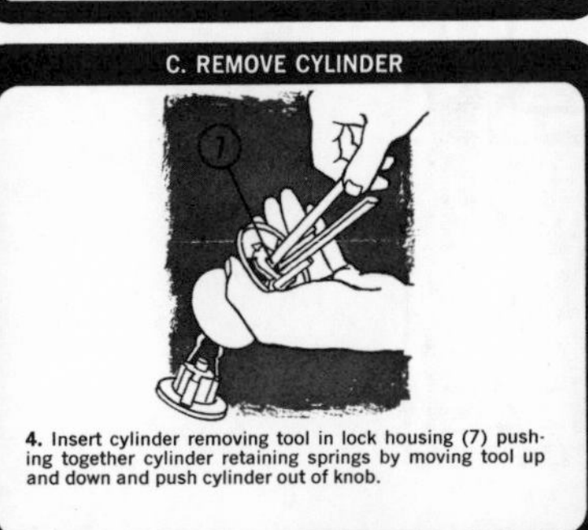

4. Insert cylinder removing tool in lock housing (7) pushing together cylinder retaining springs by moving tool up and down and push cylinder out of knob.

D. REPLACE CYLINDER

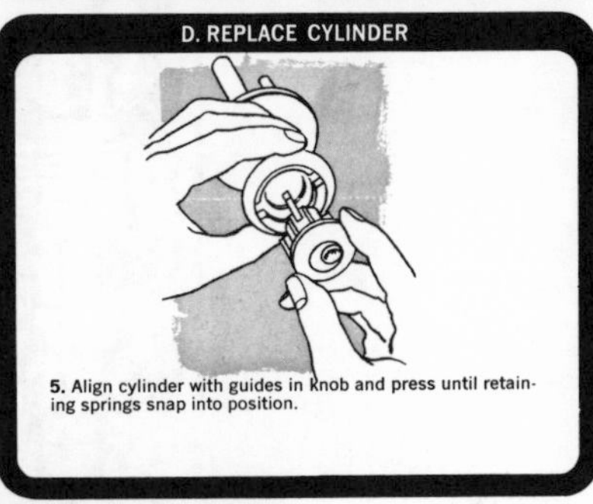

5. Align cylinder with guides in knob and press until retaining springs snap into position.

E. REPLACE SPINDLE

6. Replace round spindle by lining up boss (3) with slot (4) and push spindle until detent spring snaps into place.

52. How to remove and replace a cylinder in the Kwikset "400" line of locksets *(Courtesy Kwikset Sales and Service Co.)*

cept are still manufactured and used in England. There is the Ace, which has a key that looks a good deal like the Bramah key but which actually is a pin tumbler key fashioned in a circle. The Ace is used for pinball machines, vending machines, and the like. It is no more secure than the straight-line pin tumbler lock, but few locksmiths carry blanks, and the amateur lock pick finds it much more difficult to open than the straight-line pin tumbler lock because he so rarely encounters it. An Ace padlock is shown in Figure 53.

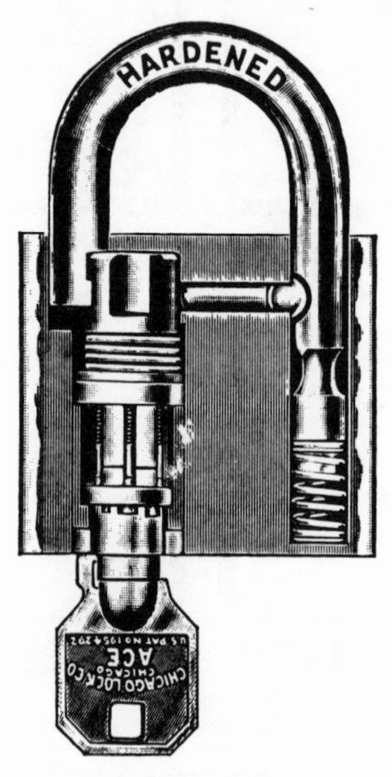

53. An Ace padlock, interior view *(Courtesy Chicago Lock Co.)*

Another lock is the combination lock, which was invented in China, although no record of its history has ever appeared in Western print. The combination lock was succeeded by the time lock, which is now our ultimate security device.

The combination lock is much too complex to be covered in a few words, but some cautionary advice is called for. Do not lock a combination safe without making certain you have its correct combination and that the lock is truly working properly. This can be done by leaving the door to the safe open and then scrambling the dials. There are very few people sufficiently skilled to open a combination lock without its combination. Although the lock you may be fooling with appears simple and has but two tumblers and a simple two-digit combination, it will withstand dozens of hours of unskilled efforts. Take no chances: Don't play with a combination lock.

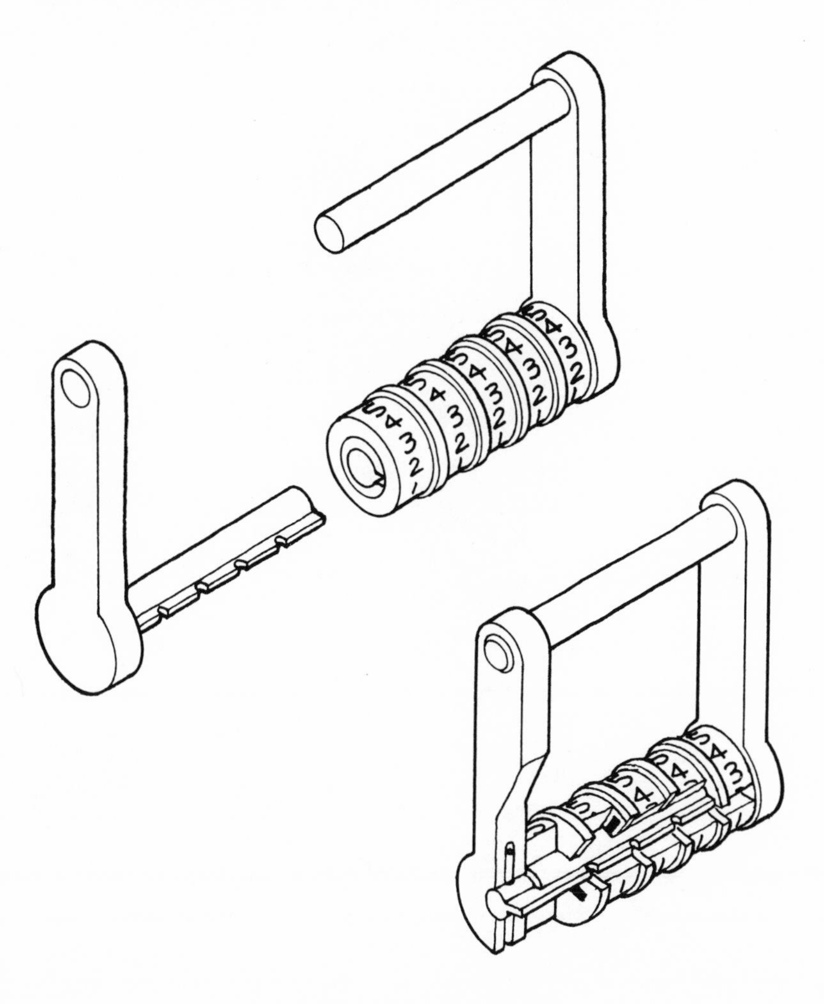

54. An old combination padlock

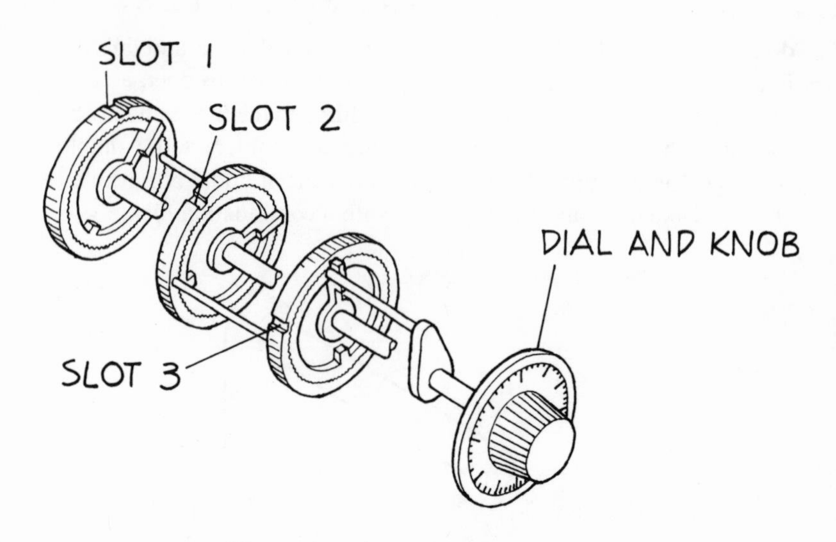

55. Inner workings of a combination lock

Chapter 8

Doors, Frames, and Their Problems

Many of the difficulties one encounters with locks often arise from door, door hardware, frame, and frame hardware changes that have occurred after the lock was installed. These changes do not affect the lock per se but do affect its operation. When conditions are extreme the lock will not close or can only be closed with great difficulty. The latch bolt refuses to hold or holds only when the door is slammed very hard. Sometimes neither the latch nor the dead bolt appears to be operative; the door can be opened at will.

DOOR FRAMES

For the past several hundred years it has been standard building practice to mount interior and exterior doors on a finished frame that is installed within a "rough" frame. The practice makes good commercial sense, for only the finished frame needs to be constructed with any degree of accuracy, and only the finished frame needs to be made of smooth, "finished" wood. The rough frame is constructed of two-by-fours, cut on the job and assembled to plus or minus a quarter-inch or more.

Figure 56 is a top view of a typical door in a wood frame. Figure 57 is a similar door in a metal frame. Like the wood frame, the metal frame is constructed at the factory. In a limited number of instances the wood frame will mount the hardware: hinges, door stop, and strike plate. When shipped, the fitted door is usually mounted

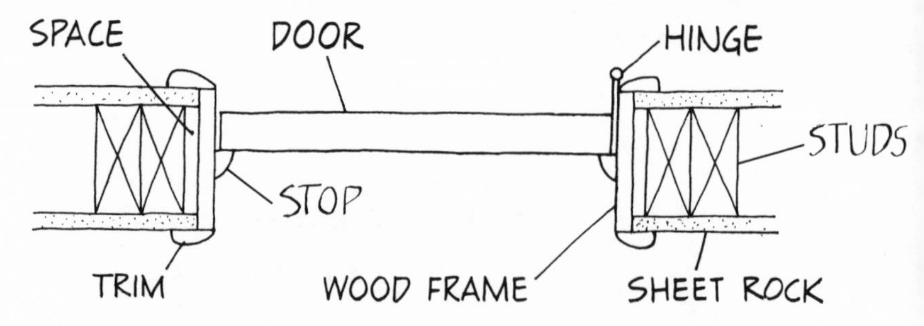

56. Top view of the construction of a wooden door frame

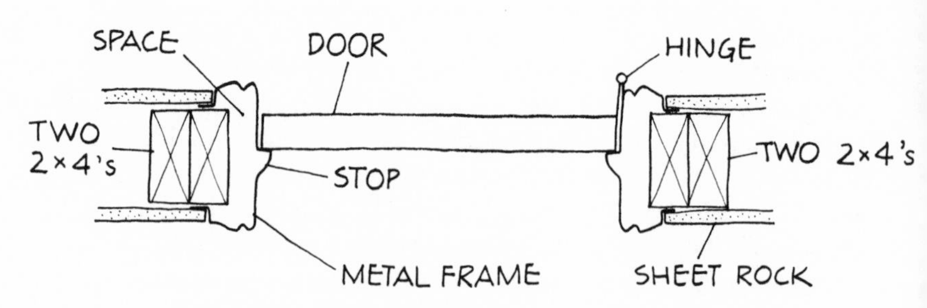

57. Top view of the construction of a metal door frame

in its frame. The metal frame is always constructed with half of the hinges or butts in place, along with the strike plate.

The wood door frame is slipped into the rough opening and nailed to the studs directly through its face. Various thicknesses of wood are used to space the frame the correct distance from the rough and to make its sides plane and vertical. Generally there may be anything from a quarter-inch to one-inch clearance between the wood frame and the supporting two-by-fours. Lengths of wood trim are then nailed to the edge of the frame to hide this opening.

Frame hardware—hinges and strike plate—are mortised into the finished frame and held there by comparatively short wood screws. Under normal conditions these screws will hold fast for dozens of years. When the screws have been permitted to loosen or when someone has closed the door on an umbrella and forced the hinges to pull the screws out, the screw holes may be destroyed. In such cases the screws are removed, and each hole is filled with glue and sawdust or partially filled with epoxy cement. The same may be done for the strike plate.

When the frame has been abused to the point where the wood has been split, the crack is cleaned and filled with epoxy cement. The crack may be clamped together until the cement has cured.

When the edges of the mortise have been torn up by forcing the door or using the door with loose hinges, the hinges are fastened in place and the space between them and the wood filled with epoxy cement. The same can be done with the strike plate.

Should the need or desire arise for greater strength than can be secured with short wood screws tied to the inch-thick finished pine frame, the frame is backed up with a length of board of the correct thickness (see Figure 58). The thickness of this board is critical. If too thick, it will press the door frame inward; if too thin, the screws will pull the frame into the studs, making the door opening wider. The trim must be removed before the back-up board can be inserted. When the trim is off, the need to fit the board will become obvious.

Run clearance holes through the door frame and the back-up or shim board before running a longer and heavier screw in place. In many instances it will be necessary to enlarge the holes in the hinge before a larger screw can be used. Be certain to bevel the enlarged hole so that the new screw head will lie flush. If it protrudes into the hinge, the door will not close properly.

Wood frames rarely warp and rarely settle, but they often suffer from over-painting and slamming. When too much paint has been applied to the inside of the frame and the edges of the door, particularly over the hinges, the door closes with difficulty. One may find that the door has to be pressed hard against the stop to make it close and that its swing is stiff from the midpoint on.

Excessive paint on the inside edge of the stop can be removed by scraping, but an easier way, though not as neat, is to lift the stop and reset it a distance away. This usually leaves a ragged paint line where the paint has to be broken.

Excessive paint on the hinges can be removed by scraping or by removing the hinges and soaking them in paint remover. Removing

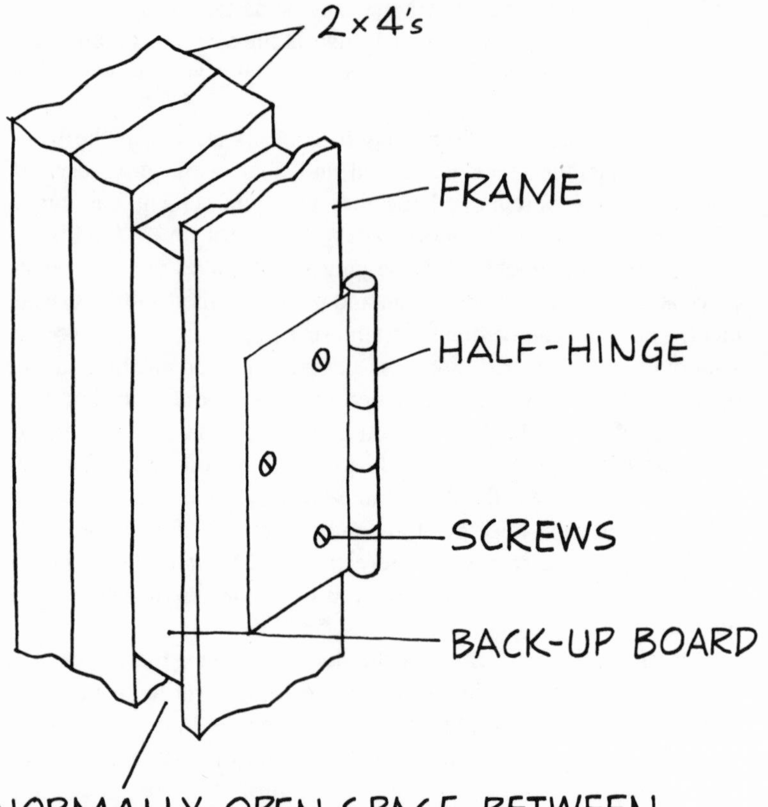

58. How the wooden door frame may be backed up with a length of board—cut to the correct thickness—so that longer than usual screws may be used and the door strengthened. The trim has to be removed to do this. Frame is shown without the sheet rock for clarity.

the hinges is not too difficult if you have a helper to hold the door. Otherwise, you will need some sort of clamp setup. If there are three hinges, you can remove and replace one hinge at a time.

Excessive paint on the hinge side of the door frame and the door can be removed by scraping or sandpapering. If you find the wood surrounding the hinges beat and torn, do not remortise. That is to say, do not insert them more deeply into the frame or door. Doing so will affect the closing of the door.

Excessive door slamming will in time loosen and move the door stop, which will result in play between the latch bolt and the hole in the strike plate. The door will rattle on windy days. Loosen the door stop and move it back into place. This is done after the door is closed and the latch bolt is in its hole.

Sometimes a door frame will change from a rectangle to a trapezoid. This occurs when one end of the house settles. The correct solution is to raise the settled end until the floor is once again level and the door frame orthogonal. The easier, but often short-lived, solution is to remove the door and cut it to an angle to match the door frame. Before going this far, check the door frame with a framing square (a large square).

Metal door frames give far less trouble, but when they do, the cure is far more difficult. Hinges and strike plates are welded in place. If they have been ripped off, they should be welded back again. Removing the door frame and replacing it means breaking plaster or sheet rock on all four adjacent walls. The alternative is to hire a man with a portable welder. If the frame has been spread a bit, it may be practical to use metal spacers beneath the hinges and longer bolts. It may be possible to use a "come along," which is a pulling mechanism strong enough to pull the parts of the door together. In some instances it may be possible to tie the "come along" to the hinges. In others a hole will need to be drilled in the sheet metal frame and this used as a tie point.

DOORS

There are two types of doors: flush and panel. The better flush doors, and those generally used for exterior passage, are solid beneath their veneer skins. The interior and low-cost doors are par-

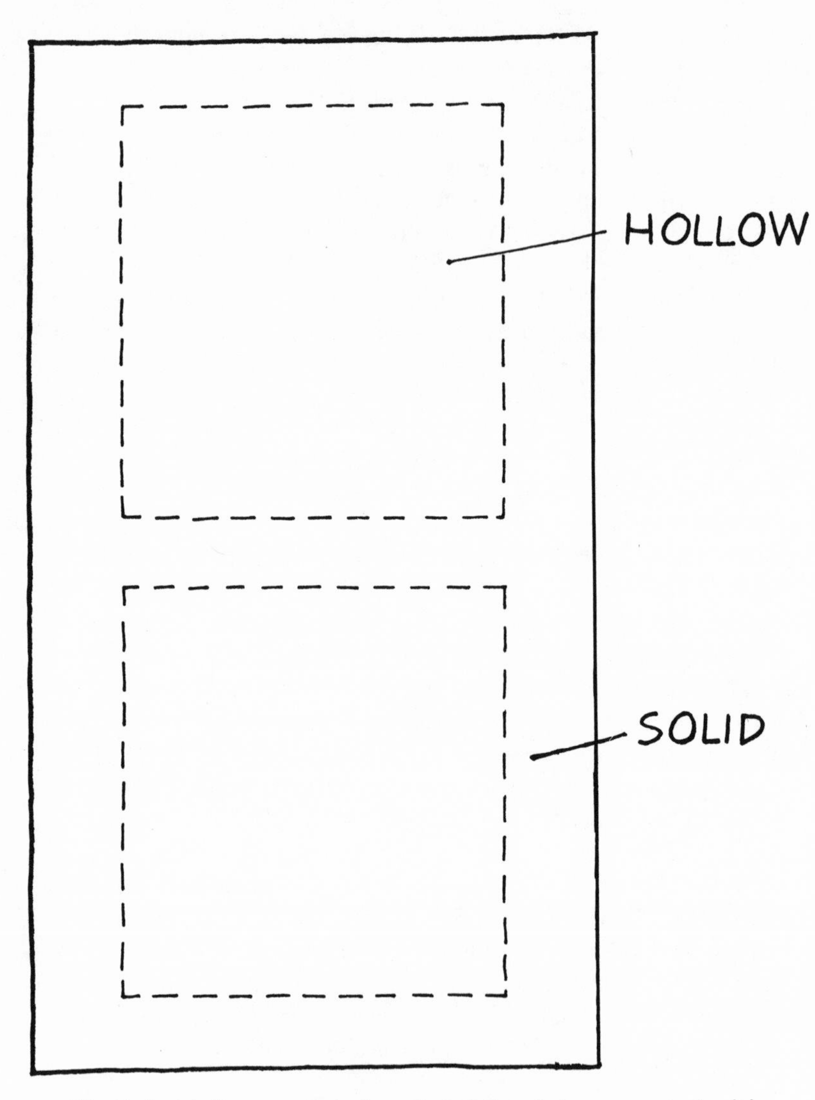

59. Typical hollow-core flush door. Dotted lines indicate extent of solid-wood interior that is capable of taking door hardware and lock.

tially hollow. The easy way to tell is to rap them with your knuckles. Figure 59 shows just about where one can expect solid wood when cutting into a hollow-core flush door.

All wood doors swell, shrink, and warp. In homes where there is no ventilation there may be as much as a half-inch difference between a door's summer width and its winter width. Obviously, since the dead bolt and latch bolt do not extend much more than a half-inch beyond the edge of the door, they may not reach the strike plate when the door shrinks in the summer.

Water-soaked doors shed their paint in large, ugly blisters. Once soaked they do not dry rapidly but will dry in time. Often these doors will warp when they dry or when they absorb moisture.

There is little point in doing much more than temporary surgery on a swollen door that is going to change its dimensions. Once it is dry, corrective steps can be taken. A dry, warped door can sometimes be straightened by use of weights. The door is removed and supported at its low points while pressure in the form of weights is used on its high spots. This takes time, possibly weeks. Generally the practical solution—if the removal of some wood will not provide a cure—is the replacement of the door.

Doors, as distinct from their hardware, will sometimes sag. This failing is almost always limited to panel doors. The simple cure is to remove the excess lumber from its bottom. If there is only a little to be taken off, place a sheet of coarse sandpaper on the floor where the door rubs and swing it back and forth. If a lot of wood must be taken off, remove the door and saw it off.

Sagging doors usually cause the dead bolt and latch bolt to miss their hole. If all is well and tight, it is better to file the strike plate than to remove it and attempt to relocate it a fraction of an inch away. There is no limit as to how much the holes in the strike plate may be enlarged, so long as the circle of metal isn't opened up.

DOOR HARDWARE

When hinges are not given a touch of grease every few years, they wear. When they do, the entire door drops. Its bottom rubs on the sill, and its dead bolt and latch bolt may fail to enter the holes in the strike plate. Some hardware shops carry washers made especially for use with hinges. It is far easier and just as good to use washers to take up the wear as it is to replace the hinges. To insert the washers, the hinge pins have to be removed. This is done by either unscrewing or forcing the lower knob from the bottom of the hinge pin. Then a drift pin or a thick nail is placed against the underside of the pin. A little hammering will remove the pin. When this is done the entire weight of the door will fall on the lower pin and hinge. Prepare for this by shimming the door up with some boards. Unless you have experience in doing this, have someone on hand to help you hold the door. If it falls, it will rip the remaining hinge from the frame.

If the hinge is badly worn and if it will not accept a washer or no washers are available for the hinge, the entire hinge has to be removed and replaced. The cost is not too high, though all-brass hinges can run several dollars. The problem is to find an exact replacement. If an exact replacement cannot be secured, great care must be used to make certain that the pivot point—that is, the pin position—is not changed. If it is changed, the door's swing will be changed. To make certain the pin position is not changed, replace the original hinge, measure its center from the door or frame, remove it, and then remeasure to locate the replacement hinge at the same place. If the new hinge is thinner, use a shim; if thicker, cut a new mortise.

Chapter 9

Master Keying

Master keying is the name given any system of locks and keys which gives one or more keys control of a greater number of locks than other keys. In a hotel, for example, the guest key opens no more than one lock. Each guest is thereby limited to his own room. The maid, on the other hand, is provided with a maid's key or a submaster key, which will gain her admittance to the rooms she is charged with and no more. The electrician or plumber, who may be called upon to make repairs in any room, might be given a master key that would admit him to all rooms, or the master key might be held by the hotel manager or his assistant, who would assume the responsibility of admitting certain individuals as necessary (and also locking nonpaying guests out). Hotels are also equipped with a grand master key with which the manager can open any door in the hotel no matter whether it is locked from the inside or the outside.

Both lever tumbler and pin tumbler locks can be master-keyed. Warded locks can also be master-keyed, but they are so simple and low on the security scale that it is doubtful if any warded locks have been master-keyed in the past hundred years.

In most instances pin tumbler locks are preferred over lever tumbler locks for master keying. One reason is that most lever tumbler master keys and systems are prepared at the factory, so that it is possible to find that the same master key is used in two proximate buildings. On the other hand, pin tumbler locks are almost always master-keyed in the field, so that the chance of a duplicate master key is very small.

Master-keyed lever tumblers cost less initially, but they require several times more space because their cases are considerably larger than pin tumbler cases. Lever tumbler maintenance is higher because many of the parts are of cast iron. When changes need to be made, the lever lock must be removed from the door. The pin tumbler cylinder is easily removed and changed. The pin tumbler has higher security, but the lever tumbler is not so easily fouled with chewing gum. This is one reason it is preferred for institutional use. Pin tumbler master-keyed installations are most often found in large hotels and public buildings. The master-keyed lever tumbler is usually found in small hotels, dormitories, steamship staterooms, and interior doors in schools.

PIN TUMBLER MASTER KEYING SYSTEMS

There are any number of different systems in use for master keying pin tumbler locks. Each system has its own advantages and disadvantages. No effort will be made to analyze them here, but they will be discussed in general terms so that an encounter with a master-keyed lock will not result in a misunderstanding of the lock's intent, leading possibly to changes and disruption of the system.

Pin tumbler locks can be master-keyed by varying keyway width (and thus the thickness of the key), varying the length of the keys, varying the number of pins used in the different locks involved, varying the bitting on the keys, and splitting the pins.

Figure 60 illustrates how keyway width can be varied to admit certain keys while excluding others. In this simplified arrangement there are four different keyway widths. The keyhole cross-section is somewhat similar for all widths.

If we give the individual room guests keys of the largest width, their keys cannot be inserted into the next thinnest keyways, no matter how the bitting of their keys may be cut. If we give the maids, for example, one step thinner keys, their keys will enter the keyways to the guest rooms and, if we desire, their keys will open their stores closets. We can divide all possible key combinations into four key thicknesses. The holder of the thinnest key can insert his key into all the wider keyways. The holder of an intermediate-

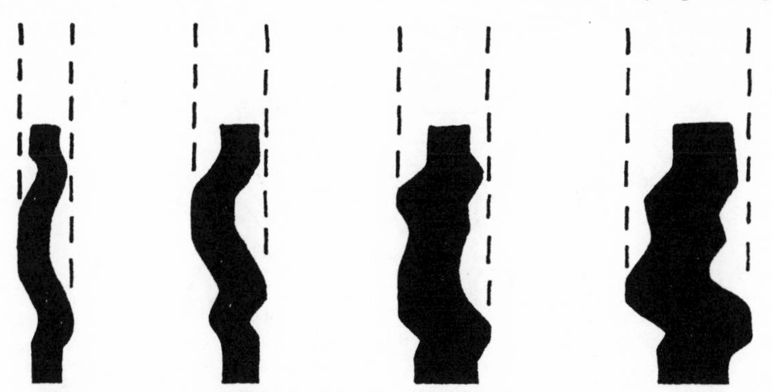

60. How varying keyway width may be used for master keying. Dimensions are exaggerated.

width key can insert his key into a wider keyway but not a thinner keyway. In this way we have four levels of authority or admittance.

If we now use the normal bitting, we can make it possible for holders of second-level keys to gain entrance to a limited number of second-level doors and a limited number of first-level doors or all first-level doors. The higher-level keys, the thinner keys, can be made to operate a greater number of doors.

Another method of master keying uses keys of different lengths and pin tumbler locks with long keyways and a limited number of pins. For example, let us use eight-pin cylinders. If the first-level locks are fitted with but four pins and these are at the front of the lock, and if the second-level locks, say the storeroom locks, are fitted with six pins, beginning with the front pins, then obviously the short keys will not reach all six pins. If the money chest, for example, is protected by an eight-pin lock, only the longest key will reach all eight pins. By the same token this long key will reach all the pins in the locks fitted with fewer pins.

How can the long key be made to operate all the shorter locks—the locks with fewer pins? One way it is done is by use of split pins. Pin lengths can be varied so that bitting height can vary from zero to nine. In other words, ten different height changes are entirely feasible and well within current manufacturing tolerances and wear experience. For simplicity, let us say that we have but two pins, and

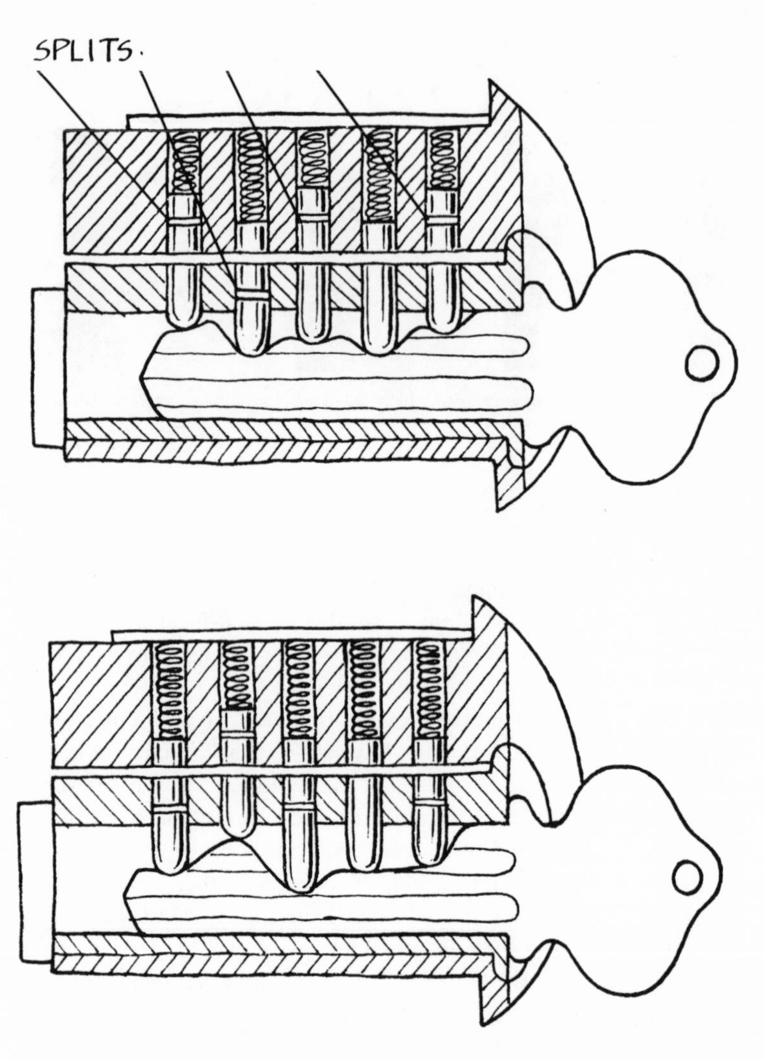

61. How split pins are used. Note the position of the "splits" in the upper lock. The inserted key will operate this lock because the pins meet or divide at the shear line. In the lower figure a differently cut key is used. Note the position of the splits. This key, though bitted differently, will also open this lock.

that lock 1 is cut to heights of three and seven. A key cut to these heights and fitting the keyway will operate this lock. Now let us take the same pins and cut each pin approximately in half. The first pin will be split at the shear line when a two- or a three-height key is inserted. The second pin might be cut to meet the shear line with a four cut. This is illustrated in Figure 61.

The guest is given a short key cut to heights of three and seven. This fits only his own room lock.

Now the manager has the long key cut for eight pins. The two pins nearest the bow are cut for two and four. His key will fit the guest's keyway and actuate the two pins at the split points. The extra key length does not actuate any pins.

By the same token, there could be a group of three-pin locks. The first split from the front of the lock is cut to two, the second to four, and, let us say, the next to three. The master key could have the same cuts. In other words, the master key could be cut to operate all the locks at the splits. The two-pin keys would be limited to the two-pin locks; the three to the three-pin locks, and all further limited by individual bitting. In the middle range, the floor manager might be given a key that has three cuts, which meet the two-four-three pattern. This could be carried on up to seven pins for the next-to-master key.

Obviously we can have a great many individual room keys, keys that open all doors on one floor, keys that open all doors on several floors but in one section of the building, keys that open all doors in several sections, and a key that opens all doors in all places.

Of course, as master keys multiply and more and more pins are cut in more and more places, lock-maintenance problems increase because tolerances become closer, and security falls. Lock picking becomes easier, and repairs become more difficult.

Chapter 10

Gaining Entry without a Key

Few heroes of fiction, movies, or television have been stopped for more than a nonce or two by a locked door. A flick of a bent hatpin and the door springs open; a twist of the wrist and the safe's tumblers fall into place. Fortunately for us dwellers this side of the magic curtain, picking locks is not all that easy. Even a simple lock requires knowledge of its inner workings and some skill. More complicated locks require more knowledge, more skill, and often considerably greater effort and time. And there is one lock, the Medeco, that purportedly has never been picked despite a twenty-five-thousand-dollar prize.

TRY THE EASY WAY

If you are locked out and it is an emergency, call the police and a locksmith. Don't waste time trying to pick the lock yourself.

In lieu of this procedure, seek the easy way in. Try the back door, the cellar door. If you live in an apartment, try going through a neighbor's apartment to the fire escape.

Look for outside hinges—hinges with their pins within reach. Remove the lower ball or knob on the pins by unscrewing or pressure. Then drive the pins up and out.

Do what a crook would do. Wrap a cloth around your hand, turn your head away, close your eyes, and break the window. Use a rock or stick if you have one. Reach in carefully.

Interior doors can be freed with an auto jack. Place the jack horizontally across the door frame and expand it. To save the trim and speed the job, remove the trim from the lock side of the door and remove any spacers that may be present between the door frame and the rough framing. This also makes it easier for the frame to bend.

LOIDING DOORS

When a spring latch bolt holds the door in place and the angle of the latch is toward you, the latch can be sprung with a flexible length of celluloid or metal. This would be the case if you were on the outside and the door closed toward you or if you were on the inside and the door swung inward.

Take a length of plastic sheet, light metal, or a slim flexible knife blade. Insert the tool between the door and the frame. Feel for the latch bolt, then press against its bevel (Figure 62). On a loose door with an accessible bevel, this can be as rapid a means of opening the door as using a key.

If the bevel is not facing you, you will need to get behind it. Make an "L"-shaped tool (Figure 63) from the material mentioned. If you can, get the "L" portion of the flat sheet behind the spring latch bolt. Then pull toward yourself. Since there may be limited space between the latch bolt and the door stop, the "hook" portion of the tool cannot be too wide.

Sometimes it is possible to slip an ice pick behind the latch bolt, wedge its end against the stop, and then swing it down to press the beveled edge of the latch bolt inward and so release the door.

PICKING DEAD BOLTS

If there is sufficient clearance between the door and its jamb, it is sometimes possible to "pick" the dead bolt with an ice pick. The point of the pick is forced against the jamb edge of the dead bolt. Then by levering the pick's side against the jamb, the bolt can be moved a little. Again, this is far from a sure-fire means of entrance. If the dead bolt is fully extended and the lock mechanism has been properly designed, the bolt mechanism will "lock" itself so that

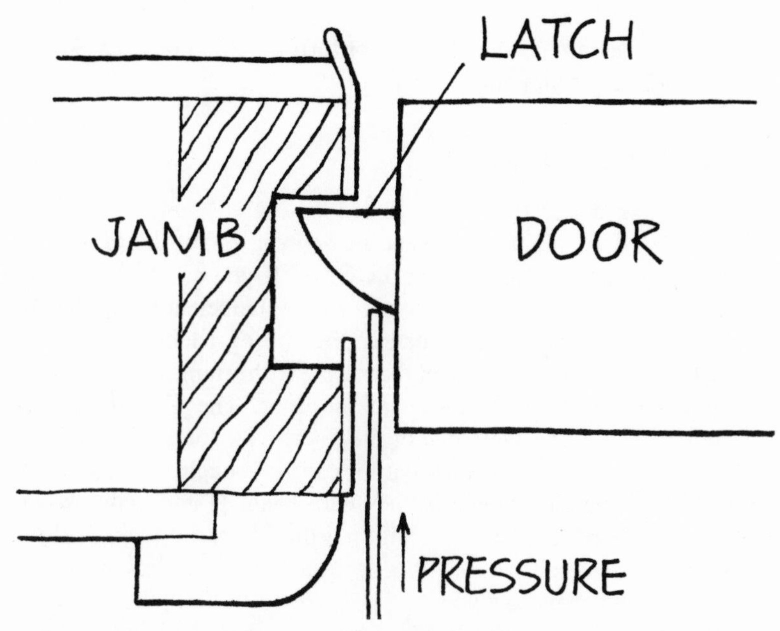

62. How a slim length of metal or plastic may be used to force the latch bolt back

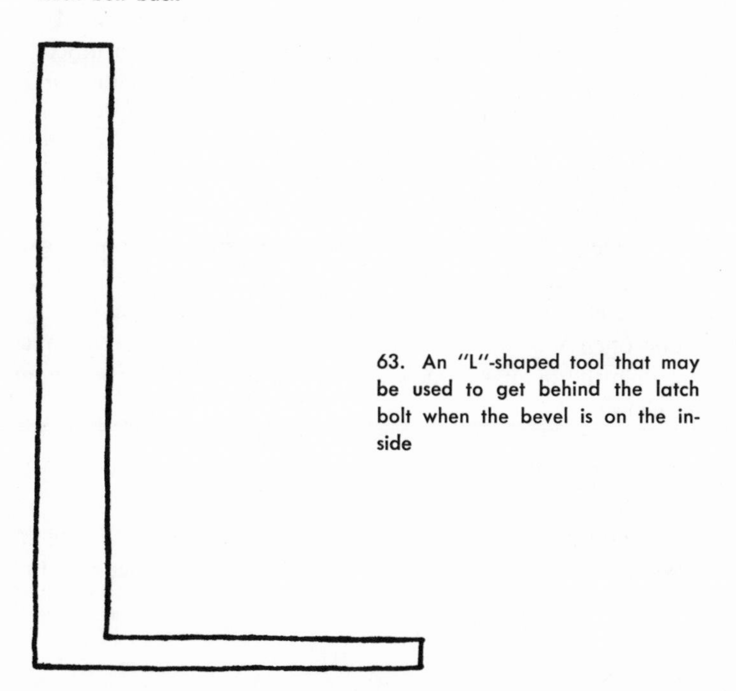

63. An "L"-shaped tool that may be used to get behind the latch bolt when the bevel is on the inside

rearward pressure on the bolt cannot turn the knob or cam. Still, it does work sometimes and is worth a try.

PICKING A SAFETY CHAIN

Assuming a situation where the door itself is unlocked, but a guard or safety chain is in place preventing entry, there are two general approaches to releasing the chain. If you can get your arm inside beyond your elbow, the chain can be unlocked with a length of scotch tape and a rubber band. One end of a length of tape is tied to one end of the rubber band. Next the tape and band are carefully brought to the inside of the door. The tape is pressed firmly against the farther edge of the door. Now the rubber band is slowly and carefully extended until it can be brought over the knob on the free or movable end of the guard chain (Figure 64). When the rubber band is released, it will pull the chain end along with itself and so release it. Usually it is necessary to keep one's fingers on the end of the chain to guide it.

If there isn't enough clearance between door and jamb to get your arm inside, try the gadget shown in Figure 65. It can be made from a length of wire taken from a clothes hanger. Obviously it will take a combination of luck and effort to get the correct arm length on the gadget so that swinging it will place the hook over the end of the guard chain. But it can be done. If you have a dentist's mirror, or any mirror fastened at an angle to the end of a stick, you will be able to see what is happening inside and your chances of success will improve immeasurably.

The practical alternative is a bolt cutter. This is a compound gear device that develops tremendous pressure. It will shear almost any piece of metal that can be placed between its jaws. If you must get the door open this way, use a large bolt cutter; it is not much thicker than the smaller models and will handle much stronger and harder chains.

OPENING WINDOWS

The standard swivel type of latch found on almost all double-hung windows (those that slide up and down) can be opened with a

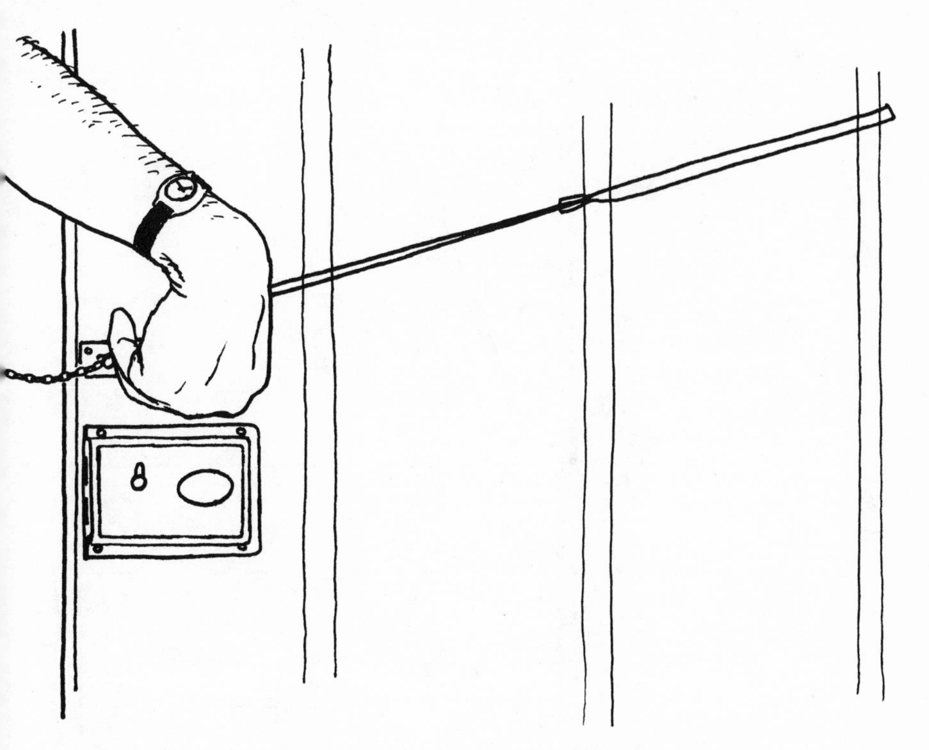

64. The old rubber-band trick—but first you've got to get your arm inside the door

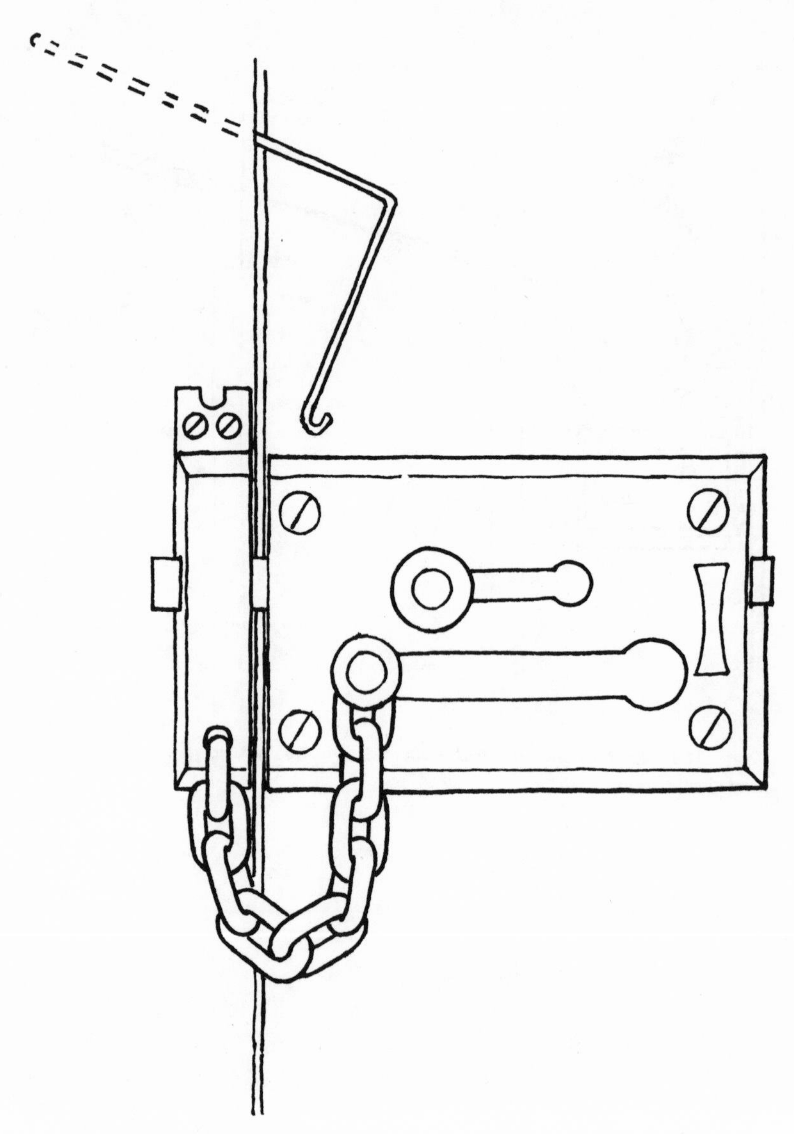

65. Using a bent length of stiff wire to move chain and then dead bolt back. This can only be done when there is a break or an absent length of stop on the door—that is, sufficient space between door and jamb to slip wire in.

knife. The point of the knife is slipped up and between the window halves (Figure 66). Then it is moved forward until it catches the latch. Pushing on the latch will turn it. If this doesn't work, use a glass cutter to score a small semicircle near the bottom of the windowpane, nearest the latch. Tap this free, than run a screwdriver inside. Wedge it beneath the latch. Turn the screwdriver, and you will most likely pull the screws out of the wood. If you don't mind a small imperfection, the small hole can be plugged with putty or a small section of glass.

Casement windows are difficult to pick. They are locked by an upward swing or rotation of a cam. The cam is shielded by the window frame. If you have time, you can remove the putty around the

66. The old butter-knife trick. This will work if there are no other devices holding the window fast.

glass, remove the glass, and then get to the latch. The glass is most easily removed from a metal frame, because there is something solid against which one can work the chisel. Wood frames are far more delicate and must be treated carefully. Breaking a stile can ruin the entire window. If you have to get inside a wood-framed casement window quickly, throw a small rock against the glass. Then use gloves to remove the pieces and be careful when reaching inside.

The same holds true for sliding windows. Generally, however, the glass on a sliding window is larger, so it is more difficult to remove the putty and more expensive if you have to break the window.

Sliding glass doors are sometimes latched by a small swinging or rotating arm. The arm is fastened to one frame and a slot in its arm engages a bolt on the other frame. In such cases you have little to lose in trying to slip a knife between the doors and upward against the latch. If you know the latch to be weak, you can chance driving the knife upward against the latch with a hammer. Do not drive it inward. You may break both glasses and damage the frame beyond repair. You might also examine the possibility of lifting one of the doors up and out of its track with a strong knife or a screwdriver. This may or may not help. Be careful if you don't want to damage the glass. If you do succeed in freeing the bottom of the glass, carry it inward, so that when you lower it, it doesn't topple away from you.

PICKING A WARDED LOCK

A warded lock has a number of physical obstacles between the dead bolt mechanism and the unauthorized individual.

The first ward, the keyhole, is easily circumvented by using either a thin key shape or a length of wire. The second ward, the side ward or wards, can be circumvented by using a wire again or a key narrow enough to slip by them. This type of key is called a skeleton key or a pass key. Figure 67 illustrates keys of the type. Note that they are narrower in two dimensions than the standard key for that type of lock.

The skeleton key is merely inserted in the keyhole and turned.

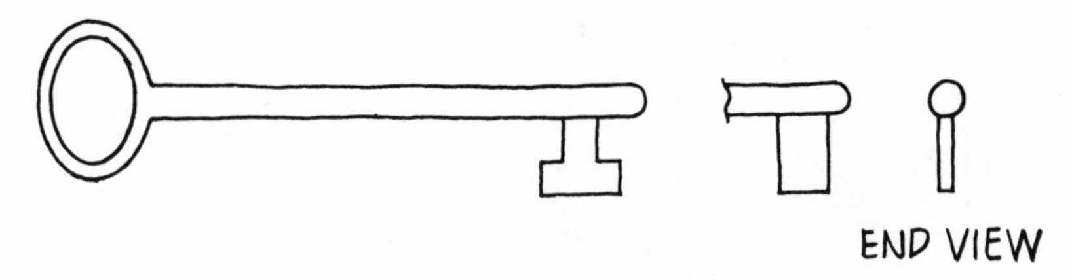

67. Sample skeleton keys

Use of a length of stiff wire or a slim allen wrench (an "L"-shaped tool) requires more thought. One must insert the short end of the picking tool into the keyhole and then feel for the lever that operates the dead bolt. Here is where intimate knowledge of the lock comes into play. The inside of the lock must be as an open book. One must see the various parts and how they are related.

What direction to turn the false key can to some extent be judged by the position of the dead bolt in relation to the keyhole. It sometimes helps to use one hand as a support for the shoulder of the false key or pick, while the other hand is used to turn the pick (Figure 68). This is not too difficult; given enough time, even a beginner can pick a warded lock.

PICKING A LEVER TUMBLER LOCK

A lever tumbler lock may be picked with two or more lengths of stiff wire with short bends on their ends. One length of wire is used to locate the dead bolt. When this is done a constant pressure is applied in a direction that will withdraw the bolt from the jamb. The other wire is then used to gently lift each tumbler up until it is in line with the post. This is done by feel. Assuming that some play has developed through the passage of time or an imperfection in manufacture—which obviously manufacturers strive to prevent—it may be possible to know when the gate on the lever is opposite the post. At that instant a little give in the dead bolt will be felt, and the lever may remain in position by virtue of the minute degree of entrance by the post into the gate. This has to be done for each lever. If the edges of the gate and the post have been serrated, the picker will be misled into believing the gate has been caught by the post,

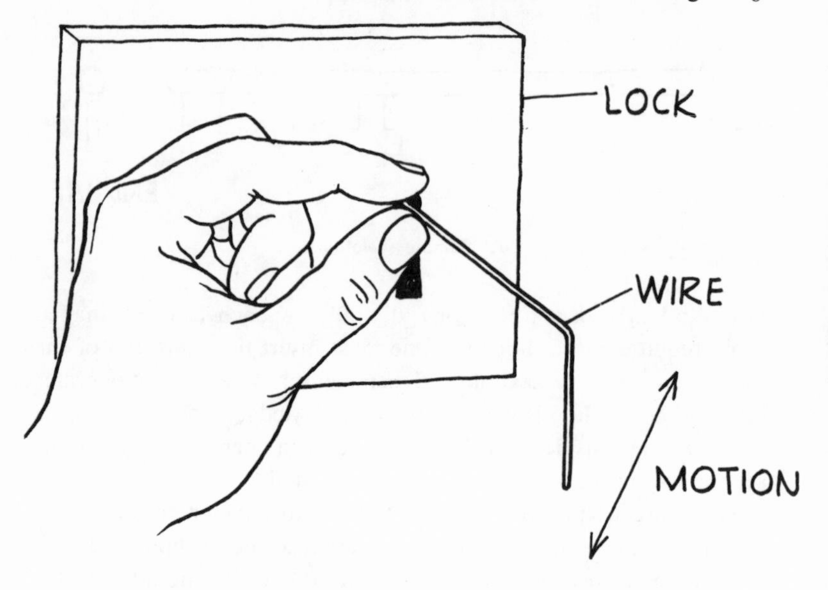

68. How a length of wire is used to pick a warded lock

when actually it is hung up on a sharp projection. As can be imagined, it isn't easy.

PICKING A DISK TUMBLER LOCK

A disk tumbler lock may be picked with two straight lengths of stiff wire or flat springs held on edge. As indicated in Chapter 5, some of the disks may be low and some may be high—assuming the worst, a double-bitted lock.

The two wires are used to perform two tasks simultaneously. One is to gently raise (and lower) the disks; the other task is to hold the sides of the wires against the sides of the disks so that a rotational pressure is applied.

The rotational pressure is necessary to "catch" the projections as they are eased into their correct position. Sometimes it is possible to rake the ends of the wires across the inside of the disks and hit the proper positions by luck and repetition. The rotational pressure, kept constant, will then open the lock.

Another approach is to use one wire or length of spring to provide the rotational pressure, while the other is used to test each disk in turn in an attempt to feel the correct position. Sometimes a small bent end on the wire helps.

CRACKING A PIN TUMBLER LOCK

With luck a pin tumbler lock may sometimes be opened in place by cracking. A tension wrench or a key that has been filed way down so that it doesn't touch the pins is inserted in the lock. The wrench or key is turned to produce a rotational tension on the lock. Then the face of the cylinder is struck a sharp blow with a plastic hammer or a mallet. If the first few strikes do not produce results, the direction of tension is reversed. If a dozen or so strikes do not crack the cylinder open, try something else.

PICKING A PIN TUMBLER LOCK

To pick a pin tumbler lock, a tension wrench is inserted in the keyway. One hand is used to maintain a turning action by means of the wrench on the cylinder's plug. Then the picking tool is inserted.

There are thousands of different shapes for picking-tool ends. Each lock pick believes that certain shapes and configurations are best for certain locks. There appears to be no consensus. However, beginners are advised to use a pick such as is shown in Figure 69. It can be bent from a length of moderate-gauge piano wire. There should be one high spot for each pin in the lock; no harm if there is

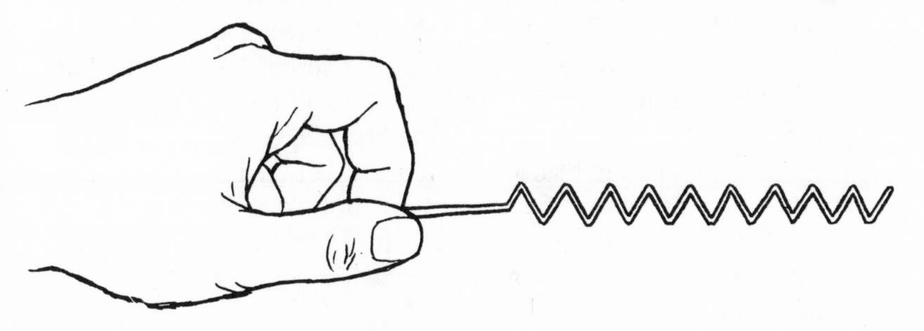

69. A pick suitable for a beginner

more, but best results can be obtained when the distance between the high spots equals the distance between pins. Designed this way, all pins are moved simultaneously. The pick is inserted in the keyway and "sawed" back and forth against the pins. It is hoped that at some particular instant all the pins are in their correct relation to the shear line. Since there is always tension on the plug, when this instant occurs, the plug will turn.

The "picking" gun is another tool suited to beginners. This device can be purchased from locksmith-supply houses. Its handle is somewhat like a gun and a trigger. Its working end consists of a slender steel wire that is inserted into the keyway and held up against the pins. When the trigger is pulled the spring vibrates and slaps the pins upward. The trigger is repeatedly pulled until the pins move into the correct position. The plug is then able to turn.

In theory, both the beginner's pick and the gun pick will eventually open any lock, although "eventually" can turn out to be a very long time. On occasion, changing the direction of the tension works. Sometimes moving the gun back and forth to vary its rhythm does the trick, as a change in rhythm, or period, changes the shape of the spring at a given moment. Sometimes it helps to shift to a different beginner's pick—one with a thicker wire or different bends.

Experienced lock picks prefer spoon-shaped tools. The technique with a spoon or a similar one-pin-at-a-time pick is different from that described. Tension must be maintained on the plug, but instead of attempting to juggle all the pins at one time in hopes of positioning them correctly for an instant, the single-pin pick is used to lift each pin individually. This is a matter of touch, and professionals on both sides of the law prefer tools without handles.

The method is to lift one pin until you can feel that pin reach the shear line, then move to the next pin, and so on all down the line. Obviously, in some cylinders, the rear pin will not remain in place while you work your way down to the first pin, or vice versa. When there is a ball bearing beneath the pin and between it and the pick, the task is compounded. Sometimes, when the cylinder is worn, the ball bearing will move to one side as it is lifted. In some very old locks of this type, a time may come when even the correct key won't work.

Generally the professional locksmith will not spend over two minutes attempting to pick a lock. It is less costly for him to remove the lock by other means and replace it.

FORCING A PIN TUMBLER LOCK

When it is impossible to pick a pin tumbler lock, and even the best lock pick cannot pick every lock he meets (sometimes a lock will respond to picking on one day but remain obdurate the following day), forcing methods are used.

A large Stillson wrench can be used to force the set screws to give way if the wrench can be applied to the rim of the cylinder. If you want to save the escutcheon behind the rim from scratches, cover it with adhesive tape. Get a good bite on the lock, and pull. If you have a two- or three-foot wrench, the set screws should give way, and the lock can be unscrewed.

If the lock is protected by a flange—a loose rim of metal that keeps the wrench's teeth away but itself turns—the job becomes more difficult. You can try to get the point of an old screwdriver under the flange and break it loose that way. You can try to drill through its edge and then break it off, or you can laboriously roll the edge of the flange back with a cold chisel until you can grip the rim of the cylinder.

Some smiths use a "nutcracker" (Figure 70) to remove a pin tumbler lock they cannot pick. This looks somewhat like a giant pair of shoemaker's pliers or nippers. If its working edges can be forced behind the rim of the cylinder, it can be used to pull the cylinder out of the door and case even though a protective flange is in the way.

DRILLING A PIN TUMBLER LOCK

Drilling a pin tumbler lock free is an easier, less destructive means of gaining entry without a key. Any type of drill can be used, including an "egg beater," which is a hand-operated drill. Bit size is not critical, but the smaller the bit used, the faster and easier it is when using a hand-powered tool. In most instances a $1/16$- to $3/32$-

inch-diameter twist drill will do the job. The hole is made along the shear line, just above the tumbler pins. If the plug cannot be turned with a torsion wrench after the hole has been drilled and the bit has been removed, a slim screwdriver or length of wire is used to push the drivers above the shear line and the tumblers below it.

70. The nutcracker. Get the claws of this pincer behind the lip of the rim lock, and it will pull almost any lock free of the case and door.

GAINING ENTRY TO A CAR

Keyless entry to a car can be gained by means other than picking the lock, which may be a disk or a pin tumbler, depending on the quality of the car. One method is by releasing the inside door-locking levers, which then release the door handles.

Figure 71 shows a number of gadgets that can be fashioned from wire and steel springs and which will act as aids in getting at the door levers. No one gadget will reach every door lever, and some levers cannot be reached because there is no rubber seal—a seal that can be pushed out of the way—between the glass and the window frame.

An alternative to picking the door's locking lever is to get in through the rear trunk. In some instances the trunk is unlocked or can be picked or unlocked normally. It is then possible, on some cars, to push the rear seat forward and get into the car.

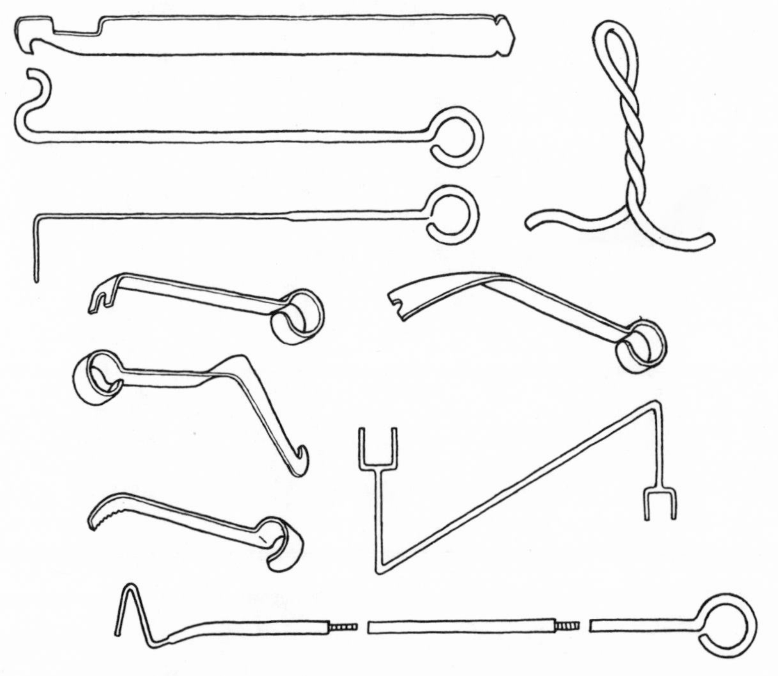

71. Some homemade tools for keyless entry into an automobile

Chapter 11

Security

The average home, be it apartment or private house, can be entered almost at will by anyone so determined. The method of entry varies, as indicated by a breakdown published in the *New York Post,* September 24, 1970:

42% (of the burglars) force open an inadequate front-door lock.
19% break back-door glass panels and unlock the door.
10% force inadequate back-door locks.
9% break windows (first or second floor).
7% break front-door glass panels and unlock the door.
6% open doors and windows.
5% break basement windows.
2% gain entry with a key.

Burglary occurs more often than any other crime. In 1968 residential losses to burglars amounted to $310,000,000, with nonresidential losses totaling another $234,000,000. In that year a burglary was committed once every 17 seconds. Automobile thefts are increasing at a rate of 100,000 per year. In 1968, 777,800 automobiles were stolen.

The cost of these burglaries cannot be measured in dollars alone. There are also fear, inconvenience, and occasionally death. When a burglar finds himself trapped by a householder, he sometimes kills his way out.

WHAT NOT TO DO

There are many things the householder can do. Even a modest expenditure of time and money can substantially increase the security of one's home. Before we consider what can and should be done, let us list those things the homeowner and apartment dweller are doing that they should *not* be doing.

Don't advertise your absence. If you are going on a wonderful cruise, fine, but wait to tell the papers about your trip until after you have returned. Crooks read too, and a statement in the press that Mr. and Mrs. Smith, of Skunk Lane, are leaving at four on Tuesday morning for a two-week stay . . . is an open invitation.

Do not forget to cancel the newspapers, milk, and mail. The post office will hold the mail for you, or you can have it delivered to your neighbor's house. Don't forget to arrange to have the grass cut, the leaves raked, or the snow shoveled.

Don't talk about your salary check, your winnings, your bonds, or what-have-you in a public place. Don't boast that your home is always open and has never been robbed. Crooks also frequent the better bars and resaurants.

Don't dash next door for a cup of coffee and leave the doors unlocked or go shopping and leave the house unlocked so the kids coming home from school can get in. Over half of all burglaries are committed in daylight.

KEEP ALERT

Now to the positive: What can be done?

Possibly the first thing to do is to learn to keep a weather eye open for crooks. Don't let anyone into your home unless he identifies himself. Don't be fooled into leaving the house by a phone call. The calls may be clever and painful. The potential thief may call and say, "Your child has been hurt; rush down to the hospital." *Check* first.

Should you see an out-of-town car roaming your streets, call the police. Any individual, whether on foot or in a vehicle, who makes it his business to pass a street several times is suspicious.

Go over your home's barriers to unlawful entry as if you yourself planned to do a job. Will your home as it stands deter a thief, slow him down, or actually invite him by the obvious absence of good locks and security devices? It is worth considering carefully.

In the main, burglars select their profession because they do not like to work. Stealing gives them the satisfaction of having beaten society, having gotten something for nothing. The thief does not wish to work and does not willingly encounter danger. If your home appears to be secure, chances are he will not probe too carefully unless he has been led to expect a large take. We can therefore deter the professional burglar by not advertising our wealth or our absence and by making our home as secure as reasonably possible.

Protection against the amateur burglar is even simpler. Much thieving is done on the spur of the moment. That is why locks are often described as devices for keeping honest people out. Pressed for money, many individuals who are normally honest will pick up and take whatever isn't "nailed down." This is the basic reason why so much stealing occurs in the low-income areas. In need of money and desperate, a man tries the door of the apartment next to his. If he can readily open it, he may well enter and grab whatever is handy and can be converted to cash. If the door resists his efforts, he will move on. He doesn't have the skill or perseverance necessary for planning or sustained effort.

Higher-income-area burglary is almost always the work of visiting crooks. They too will not spend dangerous minutes attempting to force a difficult door; they will try for easier pickings. Strengthen your defenses, and you will reduce your chances of being robbed. You can use lights, locks, mechanical barriers, guards (human and canine), and a variety of electronic safeguards and deterrents.

LIGHTING MAY DETER—OR INVITE

Much has been written about the use of light to deter crime, and it is an accepted fact that fewer crimes are committed at night in well-lit playgrounds, parks, and streets than in ill-lit or darkened areas. But light cannot be applied indiscriminately.

Recently a Westchester matron raced away on an emergency, She grabbed an overnight bag and drove off to the airport, leaving several lights burning brightly in her home. Returning a few days later, she found her house stripped to the walls; everything that could be moved had been taken. The lights had been thoughtfully turned off.

What happened to the efficacy of the lights? Nothing. But lights that do not go on and off in normal fashion indicate a tragedy or absence. We can assume that a professional burglar patrolling the area noticed that the lights remained on well past the late, late show. He waited until morning; then he and his gang backed up an innocuous truck and "moved" the house.

Continuous lighting can be used effectively in areas that are normally lit continuously. These include the driveway, the front walk, and the gate of a private home. If possible, trees and high shrubbery should be a distance from the house. The big, beautiful tree next to the building is much too good a hiding place. It produces shadows that light cannot reach. For security reasons it is best to forgo appearance and limit around-the-home plantings to small, below-the-waist bushes. Larger plantings and trees can be safely positioned a good distance away.

Normally used lights, such as those in the living room and kitchen, should be left on when one leaves the house. These lights can be made far more effective by use of a timer, however. This is a simple device that can be purchased at the larger hardware and electric shops and from mail order houses such as Allied Radio Shack, 100 N. Western Avenue, Chicago, Illinois 60680, and Lafayette Radio Electronics, 17 Union Square West, New York, New York 10003. Timers are used to turn the lights on and off at preset intervals, giving an outsider the impression that people are moving from room to room.

Lights can be effectively used by apartment-house dwellers too. Well-positioned, bright lights in the entrance hall, side yards, and rear entrances are all effective in deterring would-be thieves. Lights left on in the apartment are also useful. Often one can see an apartment light from the hall through cracks between the door and frame, and the same light can be seen from outside the building.

Although most tenants are thoroughly convinced their landlords and/or apartment managers are in league with the devil, the devil too has been known to make pacts when they cost nothing. If the cost of installing and powering additional lighting will be borne by the tenants, the landlord will have it installed. Robberies harm him almost as much as his tenants, for a reputation for robberies is not conducive to residence. For some strange reason, thieves who are successful once return to the scene of their crime time and time again. One young woman known to this writer had her New York apartment broken into and robbed three times in six months.

Lights alone cannot be depended on for home safety. Many burglars prefer to work in the daylight. They carry an insurance man's portfolio or a mechanic's tool box and blend in with the scenery. Many burglars know the local laws. In most states there is a higher penalty for illegal night entry than illegal daytime entry, and loitering in a hallway during the day is not in itself an indication of criminal intent, though it may bring a charge of criminal trespass.

SELECTING A LOCK

The preceding chapters have discussed the relative security of various locks. Anyone reading this far should be able to differentiate between a reasonably secure lock and a poor one. There are a few points, however, that have not been covered directly.

The lockset type of lock combining lock and spring latch bolt should not be used to guard an exterior door. It is too easily opened with a key. If this must be used, choose the type with an antipick latch (see Figure 79 in the Glossary).

If you already have a lockset on your front or rear door, let it be and back it up with a dead bolt lock. There are many dead bolt designs. The best was introduced by a company named Segal and is still called the Segal lock or Segal dead bolt. The bolts on this type of lock are vertical and almost impossible to pick. Two types are illustrated in figures 72 and 73. Their weakest point is their strikeplate fastening, so make certain you back up the trim and run long wood screws deeply into the studs.

If your exterior doors have panes of glass positioned so that breaking them will let the crook get to the inside dead bolt knob or latch, it is advisable to install a dual lock or dual-cylinder lock on the door. Again there are several common names for this type of a lock. It can be recognized by the presence of the normal keyhole on the outside of the door and a second keyhole on the inside of the door. There is no dead bolt knob in the ordinary sense. Once the door is locked,

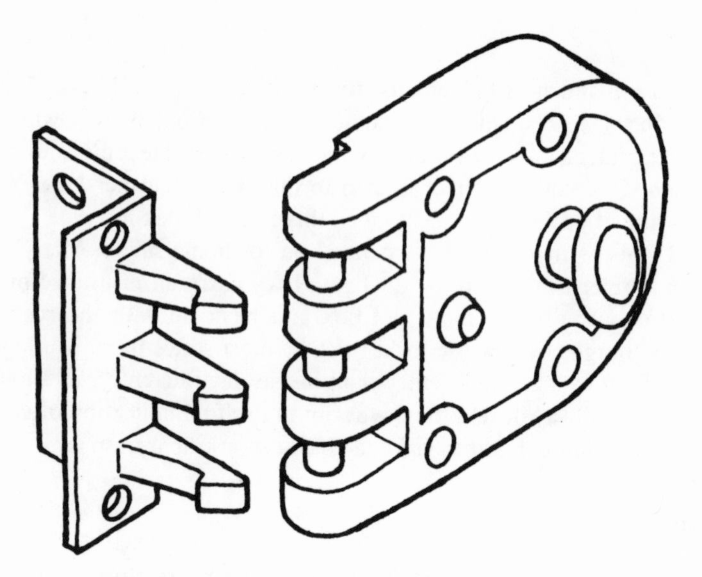

72. A Segal-type lock to be used on a solid door

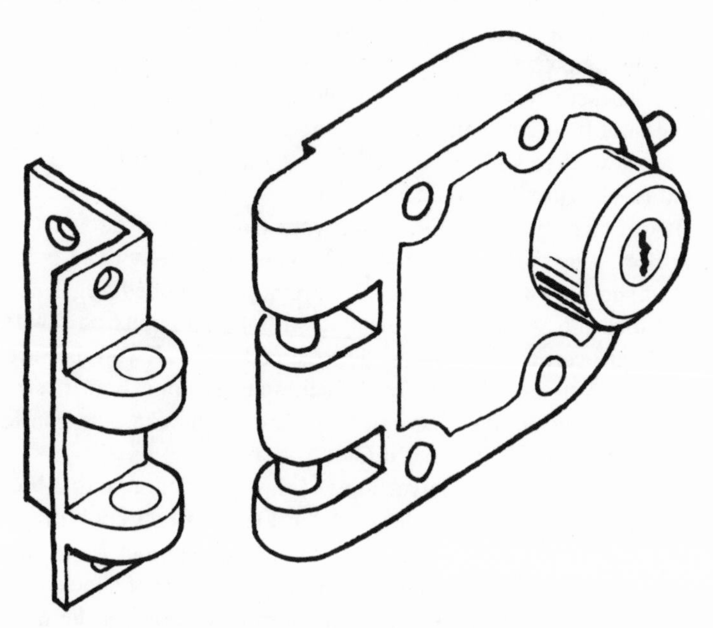

73. A similar type of lock for use on a door having one or more glass windows near the lock. Note the absence of a dead bolt knob and the presence of a second cylinder lock on the inside. To withdraw the dead bolt from either side of this lock a key is necessary.

it can only be opened with a key. Should the thief break through the glass, he still needs a key to open the door. This type of lock should not be installed when little children are left at home alone, however. They may not be wise enough or tall enough to reach the keyhole, even if they do find the right key.

Front and rear doors should also be fitted with peep holes and safety chains. A combination safety chain and dead bolt is shown in Figure 74. The type of peep hole with a small hole and a lens to provide a wide view is best. Most important: It should be used. It is of no value if the family opens the door to every friendly voice requesting audience.

Old locks and worn locks, though they may be complex, are generally considered not nearly as secure as identical but new locks. A good, solid but worn lock presents little obstacle to the expert. Left alone, he can open it in under a minute. A new lock is much more difficult; there is no play, and the shear line is much thinner.

In addition, it is good practice to replace old lock cylinders with new ones because in some instances a second-rate locksmith, called upon to make a key, found it easier to remove some of the pins than to make the proper key. Also, some of the older apartment houses were designed for maid service and were fitted with master-keyed locks. Any lock that responds to two different keys must perforce be considerably less secure and easier to pick than a lock that responds to a single key. Generally it is less costly to fit a new cylinder than to have an older cylinder fitted with new pins or changed to a single key design.

Figure 75 illustrates a popular security device, frequently found on apartment doors.

Many of the better apartment houses are equipped with a front door lock for which all the tenants have keys. When new these locks offer some obstacle to the unwanted caller. When worn, and they wear quickly because of their frequent use, they will respond to the twist of a dime. Any key that can even partially enter can operate these locks. Obviously the locks must be kept in good condition if they are to serve their purpose. Incidentally, it is best that the front door lock have its own key, rather than respond to the key used by the apartment owner for his individual door.

As good as a lock may be, it is almost worthless unless it has been properly installed in a stout door that fits its frame properly. If

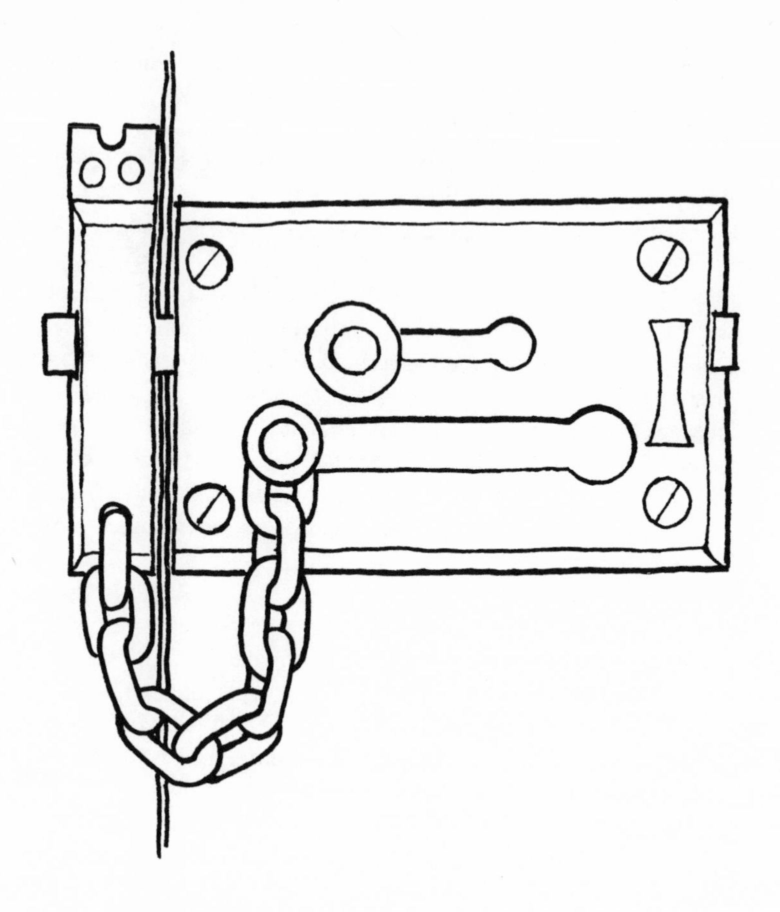

74. A combination dead bolt and safety chain device. The bolt is withdrawn and the chain removed when leaving the home. There is no external key.

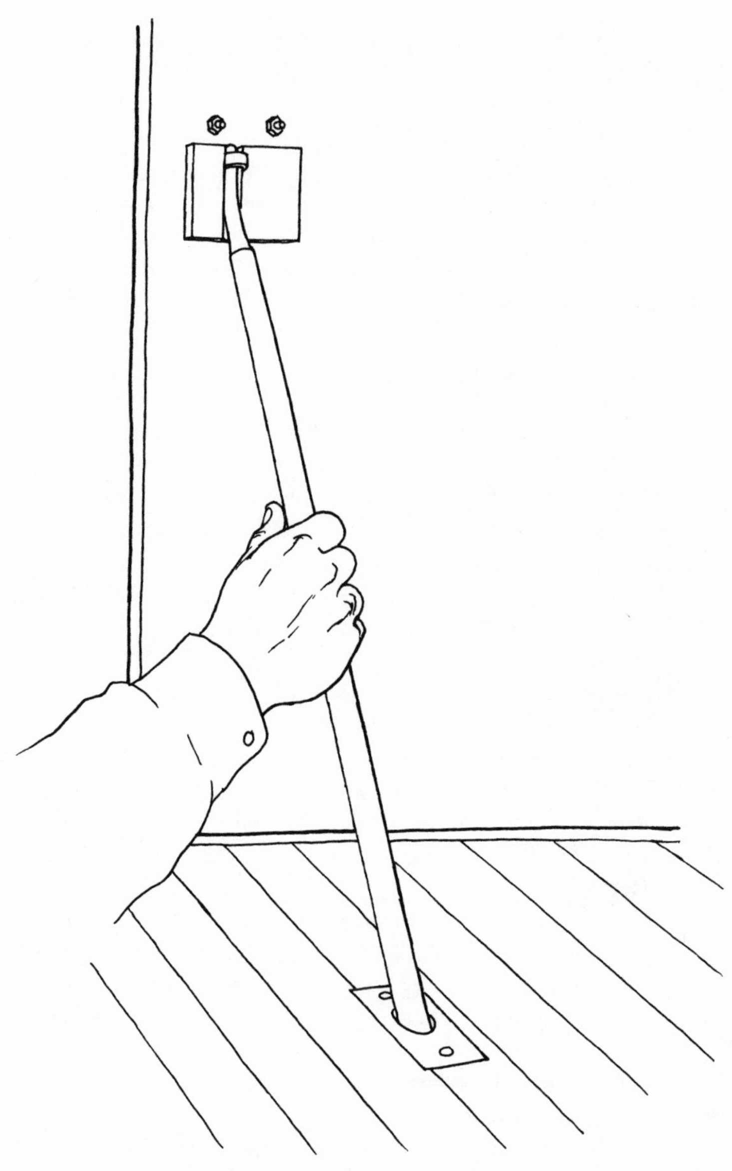

75. Door supporting device consists of a steel bar that hooks into a socket in the door and another in the floor.

there is any space between the lock side of the door and the frame, it is possible to push the bolt back, if it is of the self-locking type, by simply pressing a knife against the bevel edge. Obviously, the dead bolt type of lock is much more secure. But this too can be worked back with two sharp knives if there is enough clearance. The answer to both situations is a bolt protector. It is about six inches long with a cross-section like a "T" (Figure 76). It mounts on the door or the frame, depending on the swing of the door. The top of the "T" hides the bolt area.

OTHER FORMS OF PROTECTION

It is surprising how many householders go to considerable trouble to install a stout lock on their front door and forget their rear door and their cellar door and their garage door. An even greater number of householders fail to consider their windows as means of easy entry.

If you are planning a house, keep the first-floor windows high and small. If you already have a house, there are a number of ways to guard windows.

One can use metal grills. On dead-sash (nonmoving) windows they can be bolted to the inside frame and left in place. If that disturbs one's aesthetic sensibility, there are window-locking devices, which, though good, are not as good as a solid metal screen across a large opening. These guards come in many forms. Some have spring-powered bells that go off when the window is opened. Others are electric.

There are a number of gadgets one can make to hold windows fast. The simplest is a hole drilled through the side of the window and into the frame. A bolt is slipped through the hole. Another comprises a small hinge bolted to the window frame. At night the hinge is swung under the sash, preventing it from moving. Most hardware shops and locksmith establishments carry a number of different commercial devices for locking windows. To be effective, all windows that can be reached from the ground or from a convenient porch or garage roof must be protected.

Older apartment houses are often fitted with an electric strike on the front door and a set of push buttons with tenant names. A

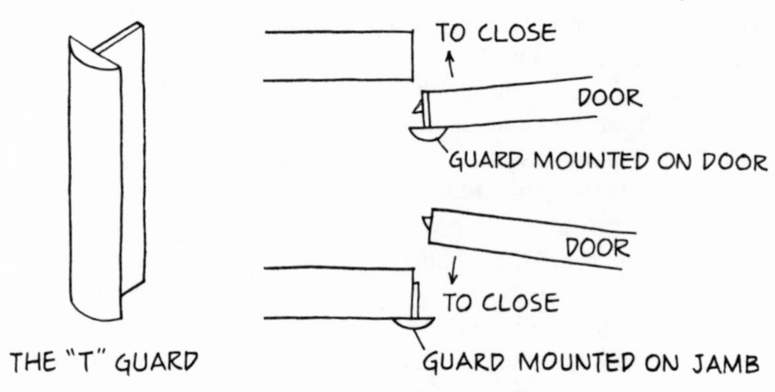

76. Use of the "T" guard to protect a latch bolt against "loiding"

would-be caller wishing admittance pushes the button of his host, who then pushes a button in his apartment that operates the strike and releases the front door. Obviously, anyone can press a button and gain entrance. As kids, this was common practice when we forgot to carry a key.

Modern apartment houses have telephone communication with the foyer. The apartment dweller can interrogate the caller and then decide to permit entry or not, but this too is easily circumvented. Most people push the button without even asking who it is; they are expecting friends. In many instances the phone system has fallen into desuetude. And if asked, the intruder can always answer, "inspector," "phone repairman," or "delivery man."

For maximum security at this point there is nothing that replaces the doorman. And in the better-designed buildings, neither front door nor rear door entry can be gained without his knowledge.

Suburban residents will also find that human guards are in many ways their most effective protection. It is possible for many communities to increase their safety at a nominal cost by hiring more police officers. The usual one routine patrol an hour can be augmented by a second squad car or even a disguised squad car that roams at random, making it that much more difficult for thieves to operate. If this is not possible, there are commercial groups that can furnish armed and trained men at a fee that is reasonable when shared by a community. In some city areas where money is scarce, the residents themselves are patrolling or standing watch at the entrance to apartment houses.

Dogs are highly useful alternatives to human guards. They have been used from time immemorial for this task. They are best when trained, as are humans, but even an untrained dog can be of assistance in alerting the household.

Some department stores use dogs. During the day the dogs may be kept on the roof. After hours they are let into the store, where they will quickly ferret out any would-be sneakthief. The gimmick here is for the burglar to stay the night in the store, then leave with the early morning crowd, the loot hidden in a store shopping bag.

Gas stations and repair shops also find dogs very helpful. These concerns often do not have the funds necessary for commercial protection. The dogs are left in the shop overnight. They are invaluable because they will bark at the slightest intrusion, and few crooks are prepared to silence a large dog or two.

Even a small dog, encouraged to bark a warning at the approach of a stranger, can be excellent protection for the homeowner. For best results, the dog should be trained. The names of schools that specialize in this work can be obtained from your veterinarian.

Lights, locks, and dogs offer considerable protection, especially when all three are working together. They should not be dismissed as means of security, but they do not approach the protection offered by humans, or that now possible with electronic devices and systems.

ELECTRONIC SECURITY

There are six basic electronic security systems: (1) the balanced bridge, (2) switches, (3) strain gauges, (4) light beams, (5) audio systems, (6) radar systems. It is believed that the military also use heat—body heat—as a detection means.

If you have looked past the diamonds in a jeweler's window, you have seen the tin-foil border along the edge of the glass. This is part of the old, but still dependable, bridge system. It comprises a closed circuit made up of lengths of foil across windows, fine wires across walls, and switches across doors and windows. The resistance of the circuit is carefully balanced in a bridge circuit. Open the circuit by breaking the foil or wire or by opening a door, and you set off an alarm. Try jumping a section of foil before you break it, and you decrease the resistance of the circuit, unbalance the bridge, and sound the alarm. The bridge system is an excellent perimeter guard system; however, installation is expensive, and it is easy to see where the glass is unprotected and to get through there.

Modern switches differ from the old in that they are not physical switches in the common sense, but proximity-sensing devices. Placed on a door or window, they sound the alarm when the door or window is moved or physically changed, but there is no physical contact between the switch and door. Generally they operate magnetically. They are easy to maintain and more difficult to spot from the outside than foil is.

Strain gauges are pressure-sensitive devices. They are usually fastened to a joist beneath the floor of each room of a house. Generally, a single unit is sensitive enough to respond to the footfalls of any person of, say, a hundred pounds wherever he may step in that room. The gauges can be adjusted not to respond to a child or a large dog. Once installed and adjusted, they and their associated equipment can be expected to hold for months at a time.

Audio devices work on what may be called the radar principle. They consist of a sender and a receiver. One sends out a supersonic signal, the other responds to it. Once set for a particular room, which may be twenty-five or more feet in any direction, any intrusion will "kick it off." What happens is that a change in the room causes a change in the reflected sound. Radar systems work the same way, but instead of using an audio signal, they use high-frequency radio signals. Neither the audio system nor the radar system is particularly applicable outdoors unless there is a considerable expanse of open area around the house or an enclosed area in which there is no movement during the hours the device is turned on.

Outdoors the audio system can be falsely triggered by the wind, a shifting tree, the passage of a large truck that reflects the sound, or the passage of a plane that makes the "wrong" sound.

The radar system too is overly sensitive outdoors. It can be triggered by a large truck that might send out the "wrong" radio signal. Neither the audio system nor the radar system will fail in its duty outdoors; they will merely be too sensitive.

Bob Dolan, vice-president of Telcoa, Inc., of Greenwich, Connecticut, a firm that has pioneered in electronic protection devices and systems, believes that the intelligent use of lights, locks, a perimeter system in the form of switches, and strain gauges probably provides the average householder with the greatest practical degree of safety consistent with reasonable cost and minimum nuisance.

Figure 77 shows a unit that combines the outputs of a stress sensor and a radar system.

So far we have mentioned devices for recognizing an intruder. These devices are in themselves invaluable but incomplete. To be complete they must warn the homeowner and others. The warning of others cannot be overemphasized.

The devices mentioned can be connected to a loud bell, which is useful in frightening the would-be intruder but doesn't help much in capturing him, or the devices can also be connected to a variety of monitoring systems: (1) They can simply send an alert signal to the local police station. (2) They can turn on a tape device which will automatically take control of the telephone line, dial the local police station (or the fire station, if that is the nature of the emergency), and relay a prerecorded message. Figure 78 shows a device of this type. (3) They can take control of the telephone line and relay the message to a monitoring service where men on twenty-four-hour duty make certain the message gets to police headquarters. The reason for the latter arrangement is that the police are often busy and may not hear the relayed message, although it can be set to repeat several times.

Additional equipment is necessary to permit the homeowner to leave the premises or enter the premises without triggering the system. Basically the method is as follows, though there are variations. The device will have two controls, each activated by a key, one indoors and one out. On leaving his home the householder turns his equipment on. He then has a preset number of minutes to leave the premises. On returning, the householder has to turn his electronic alarm system off, open the front door, get inside, and turn the second switch off before it sounds an alarm. This means a crook would need to have three keys and to use them in rapid succession.

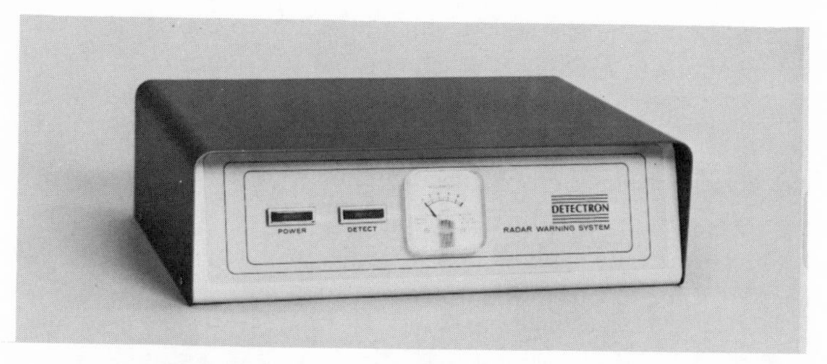

77. The Detectron control unit, which combines the outputs of a stress sensor with the output of a radar system to provide more complete intrusion protection

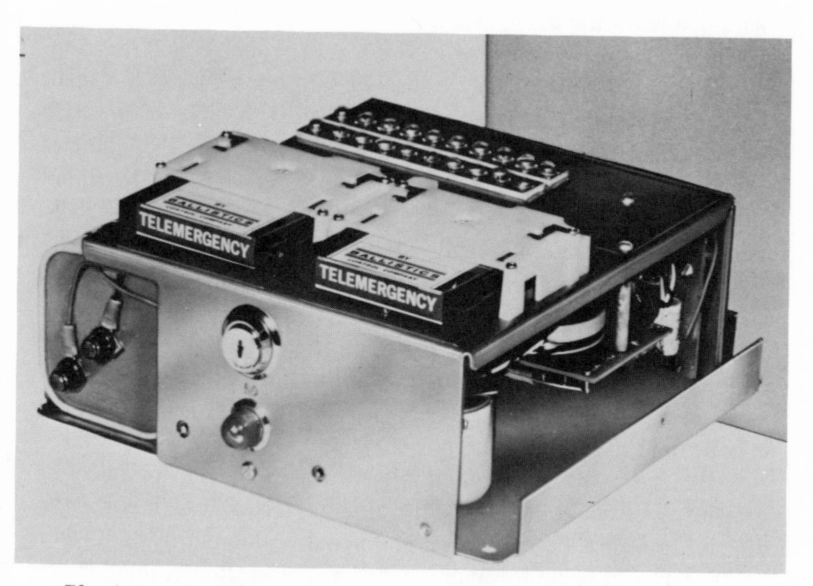

78. Automatic dialing portion of an electronic guard system. Device can be triggered by any number of means including panic buttons. The device carries two recorded messages and will take over the phone system and dial several different numbers, repeating its message twice each time someone answers. One message can be to the police department, the other directed to the fire department. *(Courtesy Ballistics Control Corp.)*

In addition, the householder can install a "panic" button at several locations in his home. When pressed, the button will sound the alarm locally and also contact the police, if so connected. Panic buttons for remote use are also manufactured for use by night watchmen. On spotting a crook, the watchman doesn't have to reach a phone; he just presses the button on his pocket transmitter.

The perimeter switches, stress sensors or strain gauges, and audio or radar sensors can all be hooked to one unit which may be connected to the automatic telephone equipment. There may also be panic buttons, horns, lights, and what-not connected to the same system.

So far we have discussed detection systems. There is also a deterrent system, supposedly the only physical deterrent available exclusive of guns that shoot automatically in the night and gas canisters that open when the wrong person enters the home.

The deterrent is a sound so loud it is unbearable. It is called Sonaguard, and it produces 138 decibels of ear-piercing noise. (A giant jet engine produces 110 decibels.) Subjected to noise of this volume, people become inoperative, just as they do under the influence of tear gas. It is believed the noise produces no damage. It cannot be stopped by ear plugs or ear muffs. In one test, a jet-engine technician, wearing ear muffs made for his work, was unable to stand up to the sound, much less function. He later said his arches hurt him unbearably.

Generally the Sonaguard is turned on to one-tenth power and left on in the guarded room. This is to warn intruders that something is amiss in that area. If they enter, any of the aforementioned devices may be used to turn it to full power.

Light circuits have not found much favor as a means of detecting prowlers. For one thing, unless the air is perfectly clear, it is not too difficult to detect the light beam and avoid walking in its path. For another, it is possible to shine a light on the detector, using a flashlight, and then break the light path with impunity. Also, the light beam cannot differentiate between dogs and intruders. If the beam is raised above the pet's head level, the intruder can easily crawl under it.

All of the electronic equipment mentioned is currently available. It must be stressed that these things are not in the "way-out" stages but are fully operative and proved. Hundreds of large companies, including banks and offices, have installed this equipment. The equipment can be purchased from the manufacturers or from distributors such as the Telcoa Company, in Greenwich, Connecticut. Local security organizations are listed in the telephone directory under "Security," or you may obtain their names from local locksmiths.

Simpler electronic equipment, including light-beam devices, can be purchased from radio-supply houses and local radio-parts houses. Their equipment costs much less but generally provides considerably less dependable protection. The better devices are equipped with back-up storage batteries that can power the alarm foı months on end should the power line fail. Some types are also wired to sound the alarm should the power line fail or be cut by a thief who thinks that shutting the power off will put the electronic devices out of commission.

ELECTRONIC AUTO-THEFT PREVENTION

Electronic guard systems for automobiles and trucks are also available. Generally, they are equipped with two locks, one inside the car and the other outside the car. On leaving the car, the owner "arms" the system by means of the inside lock. A timing circuit permits him to leave. To reenter the vehicle he has to disarm the system by means of the outside lock, which may be hidden under a fender. If it appears that all an observant crook has to do is pick the outside lock, you are mistaken. Bodily contact with the car lasting more than a preset short period of time, perhaps thirty seconds or less, will trigger the alarm. Touching the exterior lock with a metallic picking tool amounts to bodily contact.

Car and truck owners not wishing to go to the expense of an electronic guard can reduce the possibility of loss by theft through the use of any of a dozen or more supplementary car locks. To list just a few types: There is a transmission lock, called Translock, that holds a car's automatic transmission in the "park" position until it is unlocked. Another somewhat similar lock, called Parklok, does the same for stick-shift autos. Deweko Electronic Lock disconnects the ignition until it is unlocked. There is a Japanese combination lock with a ten-figure dial that also prevents ignition until opened. Krook-Lok is a gadget that locks the steering wheel to the brake pedal. Watch Dog Steering Lock, now imported from England, works on the same principle.

There are also locks that prevent unauthorized removal of your car's wheels. One device, the Buss Time-Delay Auto Protector, lets the thief drive away and then shuts the engine off. Still another system is available to protect several areas of the vehicle and the vehicle itself. Should anyone tamper with the car's lights, hood, ignition, brakes, glove compartment, doors, or trunk, two sirens go off and the engine's ignition is shut off. The car owner may also equip his car with a push button to operate the sirens should danger threaten. It's called Mobile Mini-Alarm.

For a reasonably complete list of auto-protection devices, write to Warshawsky & Co. or J. C. Whitney Co. Both are auto-parts firms located in Chicago.

A few auto-protection devices and systems are offered in local auto-parts shops and department stores. Some gas stations carry protective systems or act as agents for their manufacturers.

The standard precautionary measures advocated by the police are also highly useful in reducing car theft: Never leave the keys in the car, never leave the car unlocked, never leave the windows open—not even a crack to prevent heat buildup—and don't risk loss of the car by parking in a dark side street when a well-policed commercial parking lot is available.

Glossary

Ace lock. A lock in which the tumbler pins are arranged in a circle.

Antipick latch. A spring latch fitted with a small, parallel bar that prevents the latch from responding to external pressure from a knife blade or similar tool (see figures 79 and 80).

Armored front. A plate covering bolts or set screws holding a cylinder to its lock. These bolts are normally accessible when the door is ajar.

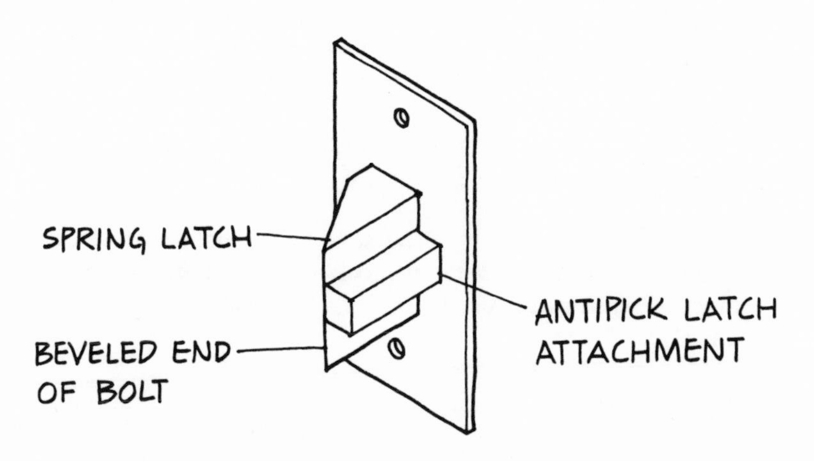

79. One type of antipick or safety latch. The safety is integral and cannot be added to standard spring latches.

80. Another type of antipick latch *(Courtesy Kwikset Sales and Service Co.)*

Back plate. This plate goes on the inside of the door. It has holes through which bolts pass and enter the rear of a rim lock, holding it in place (see Figure 90).

Backset. The horizontal distance from the edge of the door to the center of the spindle, or keyhole, if so stated (see Figure 89).

Barrel key. A key having a round stem that is hollow at its end. The hollow fits over a pin in the lock and helps keep the key orthogonal when used. Also known as a hollow-post key (see Figure 81).

Bevel. An angle. Used with doors, it designates the direction of latch bolt angle and door swing (see Figure 82).

Bicentric cylinder. A special lock having two pin tumbler cylinders. The correct key in either cylinder opens the lock. This permits master keying without reducing security (see Figure 83).

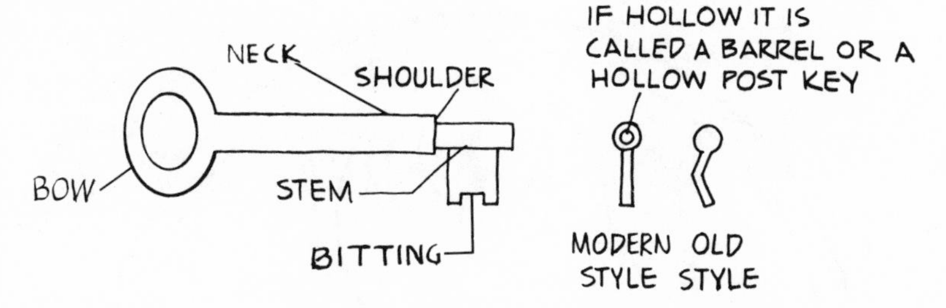

81. Typical key for a warded lock or a lever tumbler lock

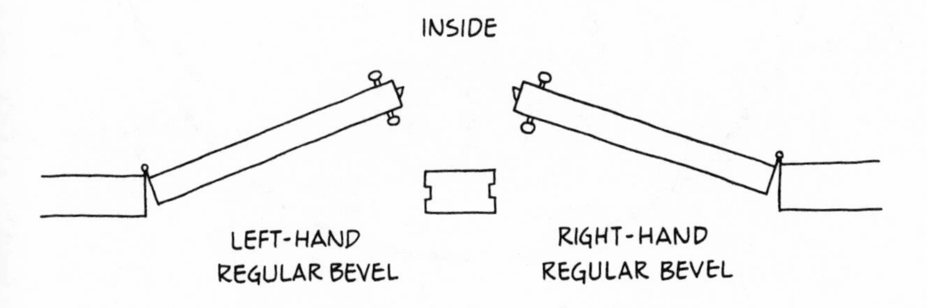

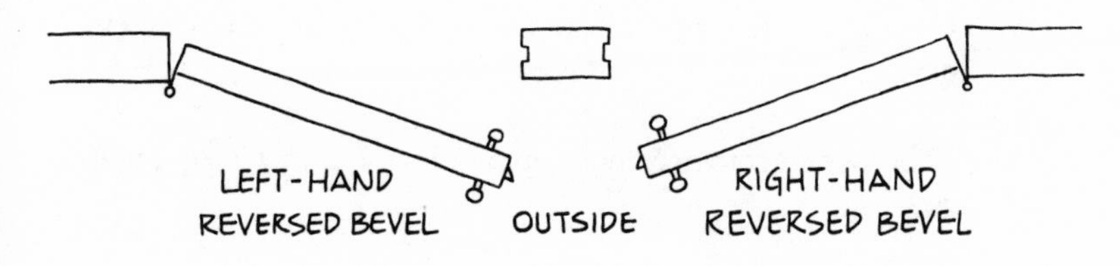

82. How door swings may be identified

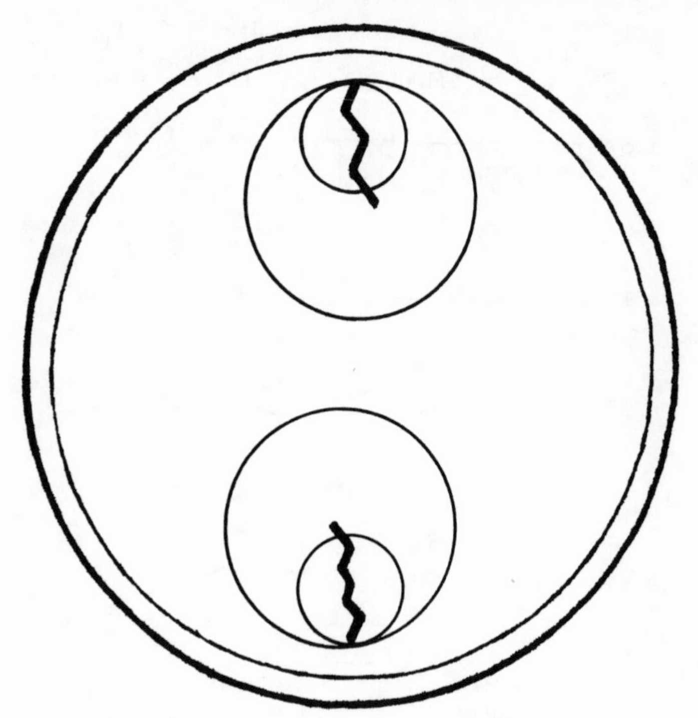

83. A bicentric cylinder lock

Bit. That part of a key which contacts the tumblers or operates the bolt. Usually it is that part of the key that is cut to fit a lock (see Figure 81).

Bitting. The arrangement of cuts on the bit of a key (see Figure 81).

Blank. A blank key that is cut or shaped as needed.

Bow. The portion of the key that is held in the hand or fingers when the key is used (see Figure 81).

Burglar-proof. A lock and door that are designed to be absolutely impregnable by any thief (without explosives and an unlimited amount of time).

Burglar-resistant. A lock and door that are capable of resisting attack by prowlers and thieves, even though they may ultimately give way to continued attack.

Cam. In locks, the cam is that portion of the lock that turns when the key is turned or when the bolt knob is turned. Unlike a tang or tail, a cam has no play and operates around a central point.

Case screw. A screw used to hold the lock's cover in place. Sometimes called a cover screw.

Case ward. A ward or obstruction generally cast integral to the case of the warded lock. The key for this lock must have a cut in its bit to match and pass the ward.

Change key. A key which fits only one lock in a building or hotel. Also known as a tenant's key, guest key, room key, and so on.

Changes. The number of different keys a particular lock design will accommodate and still differentiate between keys. The greater the number of changes, the greater the security of that lock.

Clean opening. A locksmith's term for opening a lock on a door or safe without obvious force—that is, skilled entry.

Code. Professional locksmiths have codes of ciphers and/or letters which guide them to making keys without samples. Knowing the make and serial number of a lock, the locksmith looks up his code and cuts a key accordingly.

Cog-action cam. This is a complicated cam that has an outline with more than one curve, or operates another cam or mechanism.

Collar, cylinder. Lock cylinders are designed to fit a number of door thicknesses. To accommodate a thin door, a collar is placed under the lip of the cylinder. When these are adjustable they are called spring rings.

Combination lock. A lock that does not use a physical key but requires that certain of its parts be placed in correct juxtaposition.

Generally the parts are coded with ciphers or digits and the correct word or digit series opens the lock.

Connecting bar. The bar that connects the cylinder to the lock case.

Connecting screws. The two screws that run from a plate or a portion of the lock through the door and into the lock cylinder. Removing these screws generally permits the removal of the lock's cylinder.

Core. Generally, "core" may be used to describe the inner cylinder or plug on a cylinder lock. Sometimes it is used to describe the interchangeable core cylinder found on a "Best" lock.

Corrugated key. A key stamped out of metal with a corrugated cross-section. Originally these keys were used with low-cost pin tumbler locks. Now they are restricted to use in inexpensive warded padlocks.

Crushable disk tumbler. Used in a type of tumbler lock in which disks are "crushed" to accommodate the key. They are inexpensive and provide an easy means of making a disk tumbler lock that can be matched to any key. Not used very much today.

Cuts. A general term referring to indentations made in a key blank in order to fit the key to a lock.

Cylinder. The pin tumbler mechanism, which is removable as a unit and which, when turned by the key, actuates the dead bolt.

Cylinder set screw. These are set screws, generally very long, that extend from the face of the lock into an indentation in the rim of the cylinder. When these are loosened, the cylinder can be turned and unscrewed from the case.

Dead bolt. This is a generally rectangular bolt that is not spring-operated but is moved by either the key or an inside knob. This is the part that locks the door to the door frame (see Figure 84).

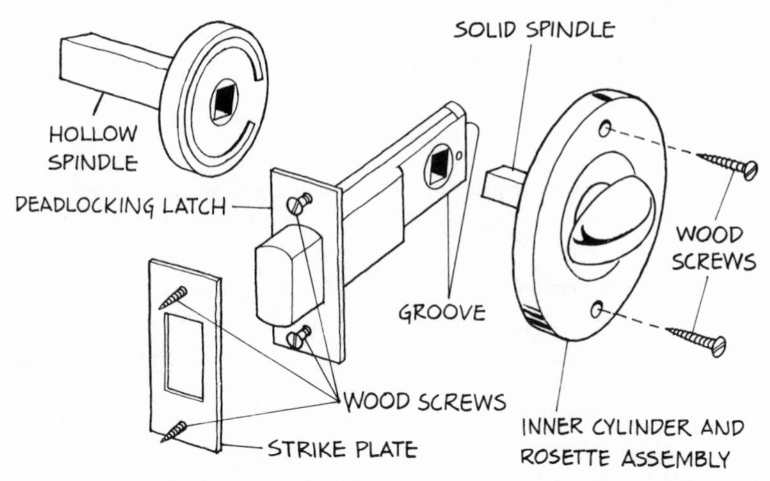

84. The dead bolt portion of a lock. This may be backed to any type of lock. *(Courtesy Weslock Co.)*

Dead latch. A second bolt parallel to and mounted alongside a spring latch. When the door has been closed and the spring latch bolt is in place, the dead latch prevents the spring latch from being pushed back into the door by any exterior picking device. Sometimes called an antipick latch (see figures 79 and 80).

Depth key. A special key that enables a locksmith to cut blank keys made for a special lock according to a code.

Derivative code. A special code which relates the lock's number with the depth of the cuts necessary to make the proper blank key work.

Disk tumbler. A circular- or oval-shaped disk with a rectangular hole and one or more projections on its edge. A number of these are used side by side in a disk tumbler lock.

Display-room key. Unlike other rooms in a hotel, a display room is fitted with a lock that has but one or two keys, which are given to the user of the room. No maids, managers, detectives, or repair men can get in.

Door check. Also called a door closer, this device consists of a heavy spring and arm coupled to an air or oil cylinder that prevents the spring from closing the door too rapidly.

Double-custody lock. A lock that can be opened only when two keys are inserted and turned. Thus it becomes impossible for either party to gain access alone.

Electric latch release. More often called an electric strike, it is an electric strike plate that releases the bolt and the door when it is energized, generally by pressing a button at a removed location.

Emergency key. This is a grand master key held by hotel and motel managers which is capable of opening any lock in the building even though the door may be locked from the inside. Supposedly it does not fit doors operated by a display-room key.

End ward. A ward on the inner case of a warded lock or a member attached to the inside of a warded lock that necessitates end bit cuts on the key.

Escutcheon. Any form or type of trim used with or on a lock. Frequently the large plate that supports the doorknob spindle and lock is called the escutcheon.

Fence. Another name for the post in a lever tumbler lock. The levers must get past this barrier to move the bolt. A fence is also the weighted plunger in a combination lock.

Following tool. The tool used to hold the pin tumblers and springs in place while a pin tumbler cylinder is being assembled or disassembled (see Figure 85).

Gate. The opening in the lever tumblers that allows them to pass the post or fence.

Genuine. Locksmith slang for a key blank made by the lock's manufacturer rather than the "counterfeits" made by small manufacturers not particularly concerned with locks.

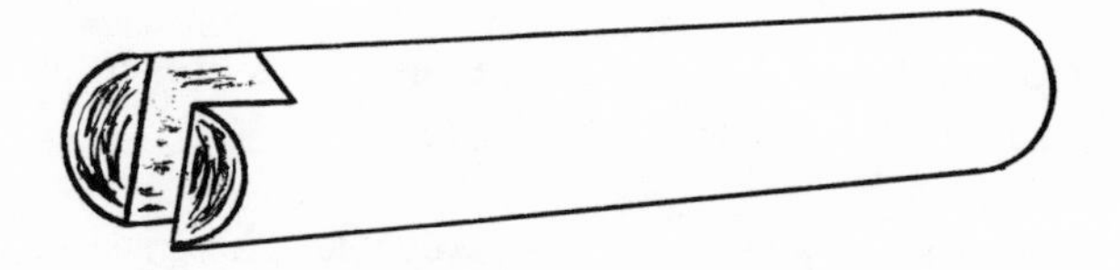

85. A homemade following tool. It is simply a short length of dowel, sanded to make it smooth, and notched at one end.

Grand master key. A master key having access to many more doors than a master key. There are grand, grand masters, and so on up the scale.

Graphite. A lubricant especially useful for the finer parts of a lock. Generally it is squirted onto the part with an air gun.

Hasp. The swinging metal band that holds a padlock to the door (see Figure 86).

Hollow-post key. A barrel key.

Jamb. The vertical members of a doorway.

Key caliper. Any small caliper suitable for measuring the height of a cut on a key.

86. A hasp is an extended hinged arm that fastens to either the door or the jamb and swings across to the other member to accept an eye, through which a peg or lock is inserted. Pictured is a hasp that engages a lock. *(Courtesy Ideal Security Hardware Corp.)*

Key check. A means of identifying a key; a tag.

Key extractor. A device used to remove key pieces broken off and remaining in the keyway (see Figure 87).

Keyhole lock. A small lock that locks a keyhole. This must be first unlocked before the main lock can be unlocked.

Keyway. The cross-section of a keyhole.

Latch. That portion of the lock that moves outward and into the door jamb to hold the door fast. Generally it is spring-loaded and is returned to the lock by turning the door knob.

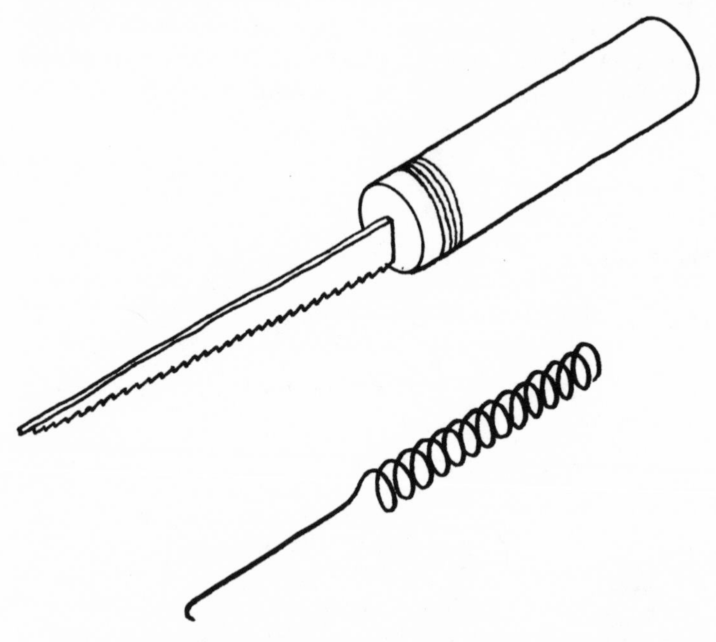

87. Two homemade broken key extractors. The upper device is made from a length of fret or keyhole saw blade. It is forced in a slot cut in a length of dowel, then wired fast. To reduce its thickness at the end, it may be ground down. The device at the bottom is a length of spring, somewhat straightened and bent into a hook. A little heat may be used to take some of the temper out of the spring.

Layout board. A length of board with a number of parallel grooves. It is used to hold pin tumbler parts in order while the smith works on the lock.

Lintel. That portion of the door frame that is overhead.

Lockset. A combination of lock, latch, and doorknobs (see Figure 88). Also, a pair of doorknobs attached to a spring latch. Also called latch set.

88. One type of lockset: a pair of knobs attached to a spring latch through a spindle. *(Courtesy Ideal Security Hardware Corp.)*

Master disk. A thin disk used when complex master keying is necessary. One or more pin tumblers are cut in half and the parts separated by a master disk, producing two positions for each pin at which the plug can be rotated.

Mortised lock. A lock that fits into the thickness of the door, as contrasted with a rim lock, which passes partly or completely through the thickness of the door (see Figure 89).

Panic bolt. A type of lock fitted with a long bar placed horizontally across the inside of a door. Light pressure on the bar releases the lock. Used in theaters, schools, and other public buildings.

Pass key. Another term for a master key or skeleton key.

Pick. Any tool or device, other than the correct key, that is used to open a lock.

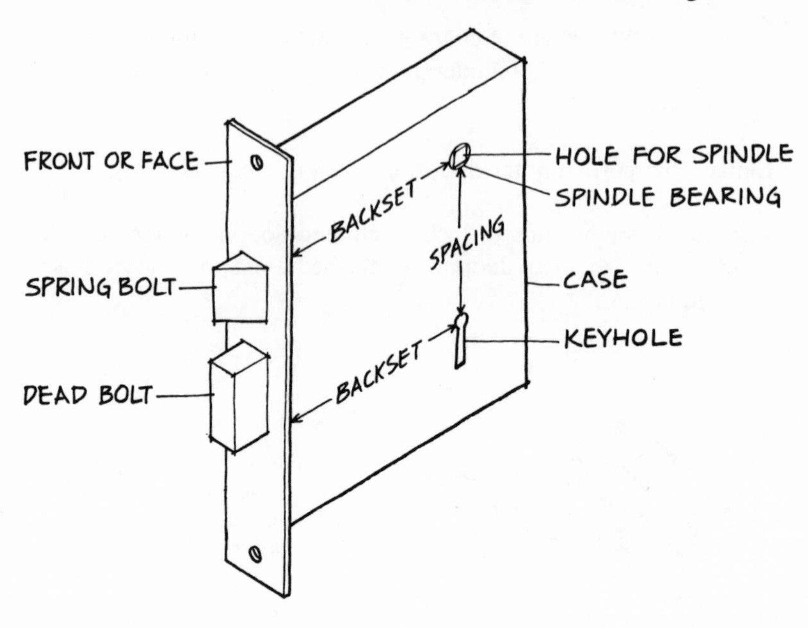

89. Major parts of a mortise lock

Plug. The inner portion of a cylinder lock; a core.

Rim Lock. A lock that passes through the thickness of a door or partway through the thickness of a door (see Figure 90).

Security. In locksmith parlance, the ability of a lock or an edifice to withstand attempts at unlawful entry.

Setup. The choice of pin drivers and tumblers which determine the key that fits a lock. These may be changed or varied in length to admit a particular key. Conversely the key may be cut to fit the setup.

Shackle. That portion of the padlock that passes through the hasp; the curved portion.

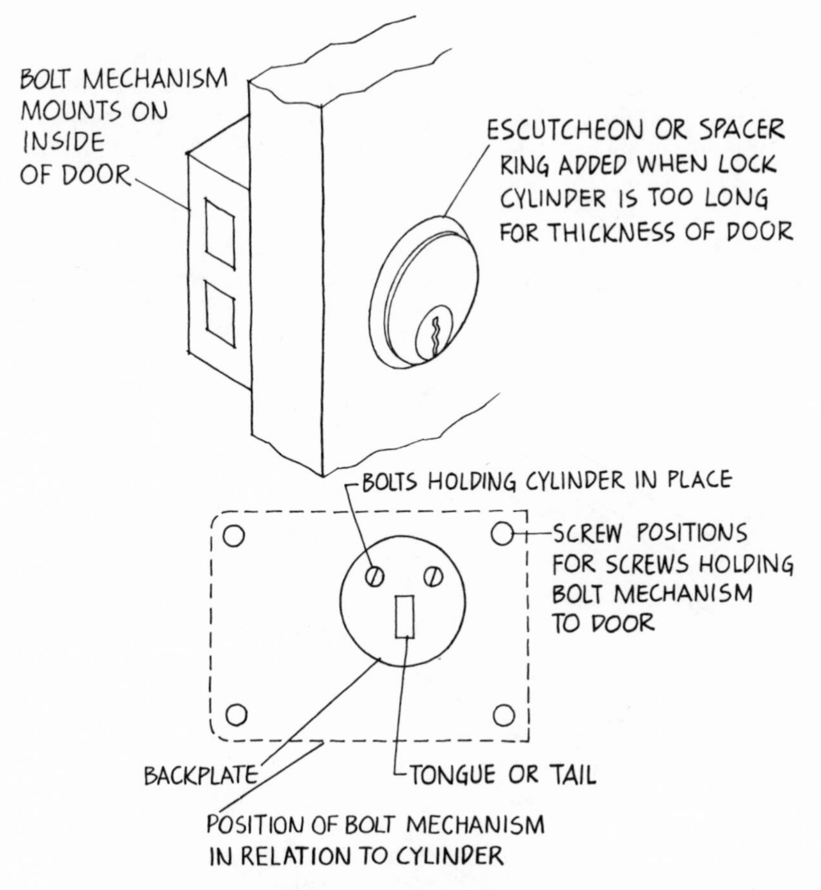

90. Inside and outside view of a rim lock

Shear line. The space between the shell and the plug of a lock cylinder. Specifically, the space between the two parts where the driver and tumbler pins meet.

Shell. The generally circular outer section of a pin tumbler lock without its inner section or plug.

Skeleton key. A warded lock key cut especially thin so that it can fit a large number of simple warded locks and lever tumbler locks.

Spacing. The distance between a lock's keyhole and its spindle hole.

Spindle. The rod which passes through the lock or door and holds the doorknobs. There are many spindle designs: swivel, split, threaded, jointed, adjustable, and single-unit spindles (see Figure 91).

Spring latch bolt. A spring loaded bolt with its external end beveled. When the door is closed, the bevel end contacts the strike

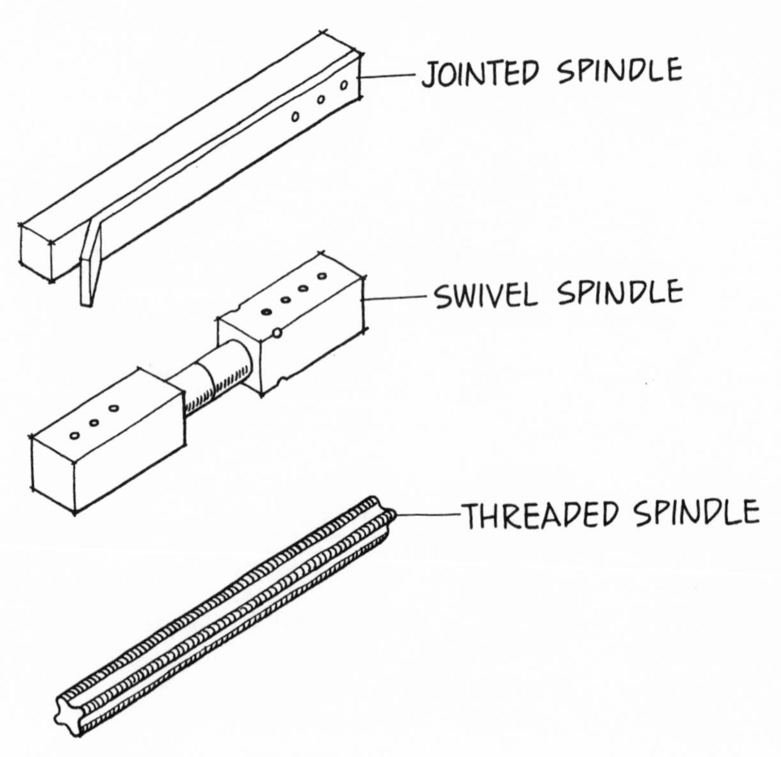

91. Three types of spindle. The length of the upper spindle can be varied by taking it apart. The length of the center spindle can be varied by unscrewing. The length of the bottom spindle is not varied. Instead the knobs are turned on the spindle and then locked in place with set screws.

plate and is pushed inward. When the door is completely closed, the spring pushes the bolt into the hole in the strike plate.

Strike, or strike plate. The metal plate attached to the door jamb which accepts the latch bolt and dead bolt (see Figure 92).

Stud. The inside-the-wall vertical piece of wood that forms the frame of the house. The studs carry the sheet rock and external walls, doors, windows, and upper floors. Studs are almost always two by four inches in cross-section.

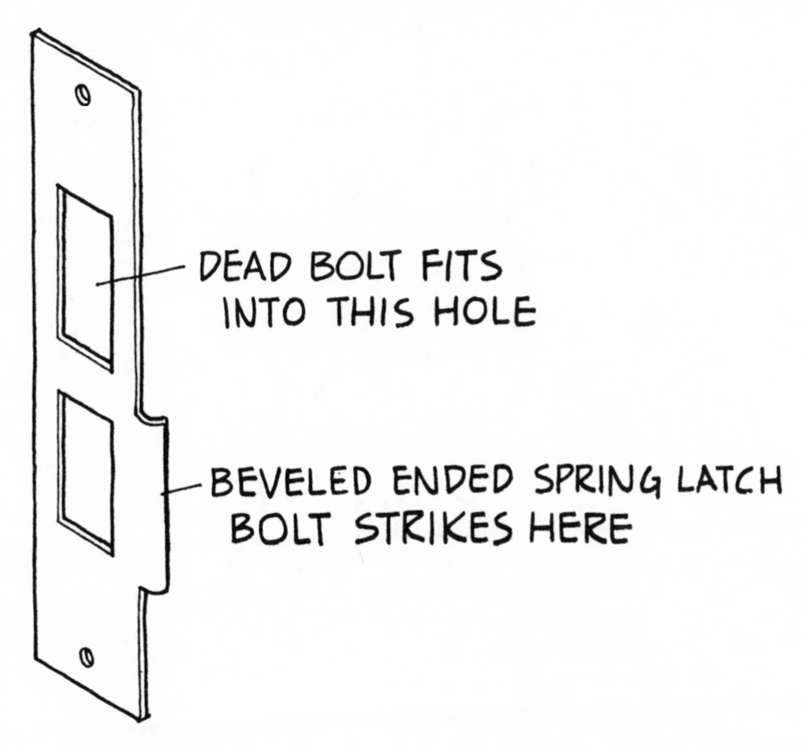

92. Typical strike plate. The spring bolt strikes the lip on the plate. This is provided to save wear on the bevel on the latch bolt. Holes for dead bolt and spring bolt are usually longer than height of bolts to reduce need for vertical accuracy when installing lock or strike plate. Increased vertical height of holes also accommodates sag in door and wear in hinges.

Tension wrench. A tool that fits into a keyway and is used to apply rotational pressure (see Figure 93).

Thimble. A tool used to hold the plug of a pin tumbler lock when working on it (see Figure 94).

Thumb knob. A lever mounted above the door handle and attached by other levers to the dead bolt or latch bolt. Depressing the thumb knob or lever withdraws the latch bolt.

Thumb turn. Sometimes called latch knob. Positioned on the inside of the door, it connects to the dead bolt. Turning the thumb knob withdraws the dead bolt.

Trim. Metal or wood parts used to hide or improve the appearance of locks, doors, and windows.

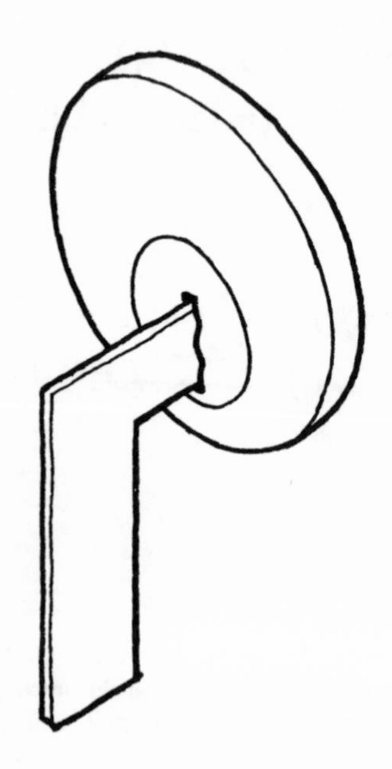

93. A tension wrench

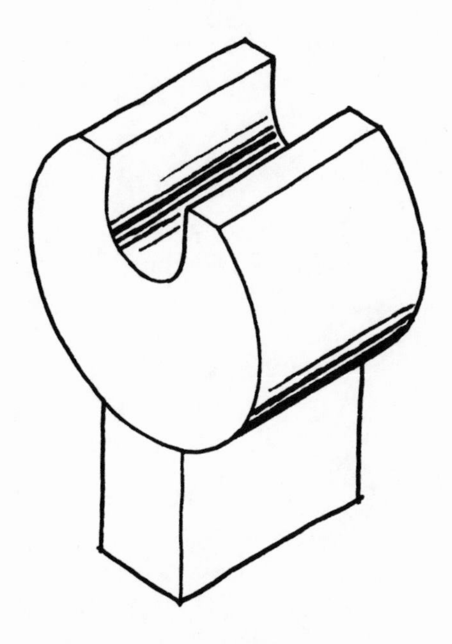

94. A thimble made from an old cylinder. The bottom is cut off with a hacksaw and file. Then a pair of holes are drilled through and a handle made of a piece of hard wood or a section of aluminum is fastened in place with bolts.

Index